Ecology of Coastal Marine Sediments

Ecology of Coastal Marine Sediments
Form, Function, and Change in the Anthropocene

Simon F. Thrush

Professor and Director, Institute of Marine Science, University of Auckland, New Zealand

Judi E. Hewitt

Principal Scientist, National Institute of Water and Atmospheric Research, Hamilton, and Professor, Department of Statistics, University of Auckland, New Zealand

Conrad A. Pilditch

Professor, School of Science, University of Waikato, New Zealand

Alf Norkko

Professor, Tvärminne Zoological Station, University of Helsinki, Finland and Guest Professor, Baltic Sea Centre, Stockholm University, Sweden

OXFORD
UNIVERSITY PRESS

OXFORD
UNIVERSITY PRESS

Great Clarendon Street, Oxford, OX2 6DP,
United Kingdom

Oxford University Press is a department of the University of Oxford.
It furthers the University's objective of excellence in research, scholarship,
and education by publishing worldwide. Oxford is a registered trade mark of
Oxford University Press in the UK and in certain other countries

First Edition published in 2021
Impression: 1

Published in the United States of America by Oxford University Press
198 Madison Avenue, New York, NY 10016, United States of America

British Library Cataloguing in Publication Data
Data available

Library of Congress Control Number: 2020945675

ISBN 978–0–19–880476–5 (hbk.)
ISBN 978–0–19–880477–2 (pbk.)

DOI: 10.1093/oso/9780198804765.001.0001

Printed and bound by
CPI Group (UK) Ltd, Croydon, CR0 4YY

Preface

As you walk over a sandflat, swim over or watch images of the seafloor you see habitats that are created by the activities of resident organisms—assuming it is not too disturbed. The organisms are involved in many interactions both with each other and with their physical and chemical environment. The habitats created can be diverse and heterogeneous, they function as complex systems. Soft sediments are partly fluid systems with very sharp geochemical boundaries and can contain diverse species with a wide range of motility at different life stages. The game rules are not necessarily the same as those that play out on competition-dominated rocky shores or in dry terrestrial systems. The structure of these seafloor ecologies (habitat forms and community composition) is, itself, important but what the seafloor ecosystems do—their processes and functioning—is profoundly important. Coastal ecosystems have been critical for humanity, providing food, other resources and some of the most expensive adjacent real estate. The level of human impacts on these functions combined with feedbacks to the delivery of ecosystem services generates both intellectual challenges and a dire need to understand and value them. The better we understand seafloor ecology and ecosystem function the more robust is the advice we can offer to policy makers and society about the consequences of change.

In this book we give you a chance to read about some of the key elements and processes involved in the structure and function of soft-sediment ecosystems. We have tried to keep the chapters short and accessible. Other authors have written books on the individual topics we cover in each chapter; our efforts are not to replace these deeper dives, but rather to provide an overview to help you see the connections between different elements of soft-sediment ecology. Soft-sediment ecology is quintessentially an interdisciplinary science with both habitats and ecological functions being the products of interactions between biological, chemical and physical processes. Thus, the capacity to connect different kinds of science is essential to advance our research. We hope the book provides you with a big-picture vision and inspires you to work on the necessary details and their implications.

The principal reason we took on the job of writing this book was because as graduate students we were all helped by John Gray's original *The Ecology of Marine Sediments* published in 1981. This was the go-to book that got us started thinking about many aspects of seafloor ecology and we continue to recommend it to our students. John also was a dear friend and inspiring colleague. We appreciated his open mindedness, enthusiasm and striving to keep ecology real and relevant by linking theory and application.

John prefaced his original book by emphasising the dismal state of soft-sediment ecology with the desire to move to a more experimental and hypothesis-testing framework. In the second edition (2009—co-authored by Mike Elliot and finished after John's death), this trend was continued with the advantage of a geographic expansion of case studies. There is still much value in consulting these earlier versions; we have not attempted an update, but rather an evolution. Now, nearly 40 years since the publication of the first edition we can point to a much more diverse and stronger science that has continued to grow in its relevance to society and inform our role as stewards of marine ecosystems.

Nevertheless, we hope this edition will demonstrate there is still a lot of research to be done in order to understand the distribution of soft-sediment communities on the seafloor and the nature and consequences of change to them. For many people, much of the seafloor is out of sight and out of mind, making it difficult for them to understand what changes over space and time are occurring, why it matters and how these changes can affect them.

Although the state of soft-sediment ecology has improved over the last 40 years, it still takes a brave individual to take on the challenge of this research. We have a much stronger experimental basis for research now. In fact we have moved beyond studying simplistic cause and effect relationships into a framework of drawing lines of evidence from multiple kinds of research conducted over different space and time scales and into a world of context dependency, interactions, networks and feedbacks. Soft-sediment ecology has also moved to address ecosystem function, which has led to an exciting interdisciplinary science and the challenges of integrating very different scientific approaches and world views. The growing societal demand on science to be relevant and connected to actions to restore degraded ecosystems, assess risks and conserve habitats and species has meant a growing transdisciplinary and social-ecological niche for soft-sediment ecology.

It would be nice to think that the next 40 years will be much more focussed on realising the deep value of these ecosystems, moving to restore them and conserve them on grand scales, rather than documenting their decline. New techniques are coming into play ranging from molecular ecology to robotics, machine learning and ocean remote sensing that will provide new ways and new scales of practical data gathering and interpretation. But you do not need expensive resources to do good ecology; understanding natural history, interactions and generating facts about how and why these ecosystems change are still vitally important to interpreting data patterns—because people need to know what it all means.

Synthesis and review are important but we cannot lose sight of the need to generate primary data. What is truly fascinating about soft sediments is not just the diversity of species but what, together with the physical and chemical conditions, these combinations of species do to make these ecosystems function and respond to change. There are many details and we can view the ecology of these systems through many lenses and much is to be learned from detailed and specific studies. However, as you focus in on the details please remain mindful of the bigger picture. This requires innovative ways of learning how to shift focus to recognise the value of different approaches and develop techniques to integrate them. In asking questions of these ecosystems, from either a fundamental or a more applied perspective, it is critical we think about scales of space, time and biological organisation and incorporate natural history, environmental context and theory.

We are increasingly aware of how humans have changed our seafloor ecology both locally and globally. Much of the growth of civilisation is linked to coasts and estuaries and many of our big cities are on or near the coast (e.g. Shanghai, New York, Manila). We continue to impact these systems both directly and indirectly over broader scales due to land–coast interactions, climate change, physical disturbance and loss of biodiversity. This means there is a real need for solutions, for understanding and monitoring change in biodiversity and ecosystem functions and, in many cases, for transforming the trajectory of ecological trends from down to up. There is lots to do—and a real need—from developing methods to studying processes, interactions and patterns.

As a point of entry this book follows the model of Gray's first edition. It is written for people starting out on a research career and a primer for established managers. It is meant to provide a starting point for new researchers to ask innovative questions and contribute to the research. We offer an overview and a sense of the connections between different fundamental concepts. Finally, let us declare our bias: although we have tried to use examples from a wide range of places, we have not delved into deep-sea sediments, but stayed closer to the coast. There are many 'world views' of soft-sediment ecology based on organisms (e.g. microbes, meiofauna or macrofauna) or processes (geochemistry, biogeochemistry, geology, physics) but this book is written from an ecological perspective of working primarily with macrofauna and their interactive role in ecosystem function. Traditionally there is a strong North Atlantic bias in soft-sediment ecological research. We have tried to find examples from further afield, being conscious of the degree of human impact in many regions. Geographical spread of research is important because without generality we can miss critical information and end up with disastrously biased perspectives. Throughout we have tried to focus on empirical research exploring and testing theory and concepts.

We have structured the book to cover five general areas. The scene is set by the first three chapters which introduce the sedimentary environment: the physico-chemical environment of the sediment; animal, plant and sediment interactions; and the seafloor as a dynamic area resulting from natural disturbances. This is followed by two practical chapters on research approaches to answering your questions. We then move on to three chapters that focus on biodiversity and communities, followed by two on ecosystem functioning. Finally we consider soft-sediment ecology and research in the Anthropocene: human impacts, climate change and restoration.

Acknowledgements

We thank Ian Sherman for the invitation to write the book. He and his staff at OUP have been helpful guides and supporters on this adventure. Writing from opposite ends of the earth, we are also grateful for and acknowledge the support from The Nottbeck Foundation and Tvärminne Zoological Station, University of Helsinki, in giving us the opportunity to get together to work on the book. Finally, we thank Jasmine Low, who has been essential in coordinating our activities in the final stages of writing and has contributed some informative figures to many chapters.

We have been lucky enough to work with many friends and colleagues that have helped, challenged and enriched our experience. This list is long but thanks to John Gray, Paul Dayton, Bob Whitlatch, Rutger Rosenberg, Roman Zajac, Sally Woodin, Dave Wethey, Erik Bonsdorff, Joanna Norkko, Paul Snelgrove, Dave Schneider, Pierre Legendre, Tom Pearson, Ragnar Elmgren, Drew Lohrer, Vonda Cummings, Joanne Ellis, Ivan Rodil, Anna Villnäs, Carolyn Lundquist, Candida Savage, Chiara Chiantore, Riccardo Cattaneo-Vietti, Silvia de Juan, Kari Ellingsen, Doug Miller, Giovanni Coco and Karin Bryan. Thanks also to our students and postdocs for commenting on drafts and more generally advancing our thinking, and their good humour in the field and in joining the dots.

Contents

PART I

The environment

CHAPTER 1

The sedimentary environment

1.1 Introduction

Our goal in this chapter is to introduce you to the sedimentary environment. Throughout the book we are going to emphasise the importance of inter-actions between physical, chemical and biological conditions. This is enormously complicated and complex, and this chapter sets the stage for much of this interaction by considering some fundamental geological, physical and chemical processes. These processes often seem simple, ordered and structured-until life starts to complicate the situation. We start with the grains that define sediments, how they are affected by hydrodynamic processes and in turn how they interact with the strong chemical gradients in marine sediments. This approach is simple but allows us to focus on how organic matter decomposition pathways vary with depth in the sediment and also how this influences the cycling of nutrients that sustain primary production. In combination, the scene set by these important physical, chemical and geological features influences not only the evolution of seabed morphology but also a wide range of ecological processes including organism dispersal and population connectivity; primary production through altered light regimes; and the release of contaminants and nutrients.

We want you to start thinking about the environmental factors that create different seafloor habitats and how they both set the scene for, and interact with, resident species and biological processes. We focus on describing why sediment grain size is one of the first variables that researchers have used to help understand variation in the structure and functioning of benthic ecosystems (Sanders, 1958, 1960). This remains a very informative variable that

we nearly always measure but it's not a simple situation where physical processes drive biology (Snelgrove & Butman, 1995). Nevertheless, sediment grain size helps us characterise the near-seabed flow regime and sedimentary environment and it is important in regulating the transfer of particles and solutes between the benthic and pelagic realms. Grain size, and more importantly how it affects the porosity and permeability of sediments, profoundly influences the microbial decomposition of organic matter. It lays the foundation for strong gradients in reactive solutes such as oxygen that influence the cycling of nutrients critical for primary producers (Glud, 2008; Middleburg, 2019).

For a book that focusses primarily on the ecology of the larger sediment-dwelling fauna, this brief primer on sediment biogeochemistry, given from a distinctively microbial perspective, may seem unnecessary. However, we believe it provides the reader with valuable context for later chapters discussing the functional roles of sediment-dwelling fauna, and how ecological interactions influence faunal diversity and ecosystem function.

1.2 Sediment grain size plus

Marine sediments are often described as sand or mud but there is a large variation in both median particle size and particle size variation. The median grain size is useful in characterising a habitat (e.g. sandflat, mudflat or gravel bed), but it is usually the variation in different components of the distribution of particle sizes that affects ecosystem processes. As the particle size varies from fine grained silts and clays (particles <63 μm diameter) to coarser-grained sands, dramatic shifts occur in

Ecology of Coastal Marine Sediments: Form, Function, and Change in the Anthropocene. Simon F. Thrush, Judi E. Hewitt, Conrad A. Pilditch and Alf Norkko, Oxford University Press (2021). © Simon F. Thrush, Judi Hewitt, Conrad Pilditch, and Alf Norkko.
DOI: 10.1093/oso/9780198804765.003.0001

densities and types of fauna found inhabiting the sediment and how they interact to influence ecosystem function. The range of particle sizes that comprise the sediment regulates the abilities of animals to construct burrows, how easily sediments are transported by waves and currents and the exchange of solutes across the sediment–water interface. Shifts in benthic diversity and ecosystem function often correlate well with grain size although the form of these relationships can vary depending on the range of sediment grain sizes included in the study. At the extremes, the finest-grained mudflats and the coarsest beach sands, the diversity will be low but important changes occur across a range of muddy sands to medium sand habitats (e.g. Douglas et al., 2018; Pratt et al., 2014).

In coastal habitats grain size varies across multiple spatial scales as a function of sediment supply from land and the near-bed hydrodynamic forces which transport and sort sediments. These processes can be in play on a regular basis associated with tidal flows or wave action, but extreme events, such as storms, can set up sediment sources and physical structures, such as sand waves, that may structure habitats for longer periods (Green & Coco, 2014; Hall, 1994; Traykovski et al., 1999). As we will see in Chapter 2 the activities of benthic organisms are also an important driver of grain size distribution. On the continental shelf, distribution reflects bathymetry where finer grains accumulate in deeper waters and coarser grains persist in an absence of fine sediment sources and/or persistent winnowing by near-bed flows. Within estuaries, variations in grain size are correlated with different habitats where muds accumulate in some areas and coarser sands in others. These basic physical settings (sediment supply and hydrodynamics) are further modified by the presence or activities of organisms (e.g. seagrass meadows and mangrove forests may trap fine particles). A detailed review of the particle grain size analysis methods for marine sediments can be found in Kenny and Sotheran (2013).

Size classes used for characterising the sedimentary environment follow the sedimentary geology Wentworth scale, which is a geometric scale that covers 32 mm, 16 mm, 8 mm, 4 mm and so on down to 0.0039 mm (Table 1.1; Leeder, 1982). For

Table 1.1 Wentworth description of sediments based on particle diameter size and the phi scale

Grain size (mm)	Phi (Ø) scale	Sediment type	
>4	<−2	Pebbles–boulders	Gravel
2	−1	Granule	
1	0	Very coarse sand	Sand
0.5	1	Coarse sand	
0.25	2	Medium sand	
0.125	3	Fine sand	
0.0625	4	Very fine sand	
0.031	>4–8	Silt	Mud
<0.0039	>8–10	Clay	

convenience these size classes are often transformed to the arithmetic phi (Ø) scale which is defined as the $-\log_2$ of the particle size in mm, so that a unit change in Ø represents a factor of two change in grain size.

Unless the system is high energy, such as a beach face exposed to waves, or extremely low energy, accumulating fine silts/clays, most coastal sediments represent a mixture of size classes reflecting the complex interactions between supply, physical transport and ecology. Thus, it is important to consider not only a median grain size but also a sorting index. This index describes the distribution of particles in a sample. Although there are a number of ways to measure sorting, the recommended index is the Inclusive Graphic Standard Deviation.

Whereas the median grain size and a sorting coefficient are sterile numbers, a critical element, depending on your questions, is either the percentage of fine particles or the percentage of coarse particles. In a sedimentary matrix these two percentages fundamentally change the sediment's properties. Thus, we frequently characterise sediment based on amount of mud (or particles <63 μm in the silt/clay fraction) according to Folk's (1954) terminology, which provides a more informative summary of the grain size distribution and physical properties (Figure 1.1).

Grain size and sorting alone are not the only sediment characteristics useful to measure (Box 1.1). The other key measures are the sediment porosity and permeability. Porosity (Ø) is a measure of the void space between grains (V_p) but is presented as a ratio

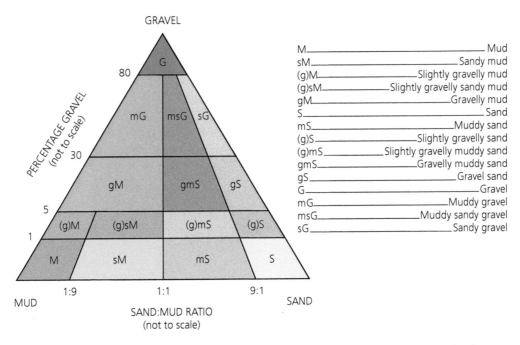

Figure 1.1 Simplified Folk classification system for sediments with mixed particle size ranges. From Figure 1, Long (2006).

in terms of the bulk volume of the sediment (V_T; i.e. $\emptyset = V_P/V_T$) with values ranging from 0 to 1. It can be easily measured gravimetrically (see Kenny & Sotheran, 2013). The relationship between grain size distribution and porosity is somewhat counter-intuitive; a sandy sediment will have a greater porosity than a silty sand because the silt particles clog the interstitial voids, but then as the silt/clay content increases so does the porosity as the water content increases (e.g. Kamann et al., 2007). Permeability (K; m^2) is an indication of how easily water can move through the pore spaces and can be measured as the rate at which water under a constant head of pressure passes through a sediment core of known diameter and length (Klute & Dirksen, 1986). Porosity and permeability are not necessarily correlated because permeability is a function of the size of the pore spaces and their interconnectivity. Muddy sediments can have a high porosity (water content) but, due to the very small pore spaces and lack of connectivity, have a low permeability. Sandy sediment on the other hand can have a high porosity and permeability.

> **Box 1.1 Sediments can also be defined as cohesive or non-cohesive**
>
> In non-cohesive sediments the individual grains can be eroded whereas cohesive sediments stick together and do not behave as separate particles. This is a property determined largely by the silt/clay fraction. Non-cohesive sediments are generally sands with a low silt/clay content (< ~10%) and are characterised by a high permeability. However, as the silt/clay content increases, filling the void spaces and ultimately binding the sediment together, due to electromagnetic forces, it creates a cohesive sediment with low permeability. The distinction between these two sediment types is an important one because it helps define the erosive behaviour when subjected to increasing currents. It also regulates the non-biologically mediated transport of solutes (such as oxygen and nutrients) across the sediment–water interface. In cohesive sediments the process is slow and dominated by molecular diffusion where solutes must follow a twisted path through the sediment matrix, whereas in permeable sediments near-bed flows and pressure variations can induce much more rapid exchanges of solutes (see section 1.4).

In many coastal settings sediment grain size is not uniform. Whether these sediment mixtures behave cohesively or non-cohesively depends on many factors including the fraction of fines; compaction and the size ratio of the small and large fractions; and, of course, ecological processes (see, e.g. Bartzke et al., 2013; Le Hir et al., 2007; Jacobs et al., 2011; Staudt et al., 2017 and references therein; Box 1). The important thing is that relatively small amounts of fine sediments added to the surface of sands can cause a shift between these two behaviours—a small amount of silt and clay can make a big difference! If we measure grain size of the near-surface sediments (0–2 cm) we may see this effect, but if we take deeper cores to assess the average grain size of the sediments we may dilute the effects of these surficial deposits. This is important because in many coastal areas fine sediment addition has changed or is changing grain size distribution (see Chapter 11).

The larger plants and animals can radically change the physical characteristics of sediments, for example, by creating faecal pellets that bind small particles into larger ones or selecting certain grain sizes to build tubes (see Chapter 2). This led to consideration about how we should actually define grain size, but measuring the size of the physical particles (involving removal of organic matter and disaggregation) gives us more consistent and comparable data. Nevertheless, it is important to remember that realised grain size distributions, porosity and permeability may be quite different than those predicted from inorganic particle size analysis.

1.3 Flow, waves and the benthic boundary layer

The characteristics of the near-bed flow play a key role in regulating solute and particle fluxes to and from the water column and thus the degree of connectivity between the benthic and pelagic environments. The interaction between flow, the sediment and organisms inhabiting the seafloor influences feeding rates of benthic organisms, supply of organic matter, resuspension potential of sediments and thus disturbance regime, recruitment and the strength of biotic interactions such as adult–juvenile

interactions. The forces generating flows on the coastal seabed vary widely in their spatial and temporal scales and include tides, waves, upwelling, storms and coastal currents. The importance of flow interactions with microbes, larger plants and animals will be covered in Chapter 2, here we will give a general overview of near-bed flows and their consequences for particle and solute fluxes. -see Nowell and Jumars (1984), Jumars (1993) and Boudreau and Jorgensen (2001) for an in-depth introduction to the physics and consequences of benthic processes.

Whether flow is likely to be turbulent or laminar can be assessed by calculating the Reynolds number (Re), a dimensionless number (i.e. it has no units so is independent of the scale of observation). It compares the factors responsible for generating turbulence (fluid speed and an appropriate length scale that the fluid is interacting with) to those that dampen it out (viscosity of the fluid). Estimating the Re for a given situation gives a first-order approximation as to how quickly mixing will occur and can also give an indication of the boundary layer characteristics (Box 1.2).

In coastal environments the thickness and structure of the benthic boundary layer (BBL) are highly dynamic, due to spatial and temporal variations in flow speed and bottom roughness. Under slow, steady currents the BBL thickness may be tens of cm but under high flows it may only extend a few cm above the seabed. Furthermore, as flow speed and/or bottom roughness increases, the viscous and diffusive layers may disappear altogether and the logarithmic turbulent layer penetrates all the way to the bed.

The steepness of the velocity gradient in the BBL determines the force exerted at the seabed which influences particle deposition and resuspension. This force is represented by the bed shear stress (τ; N m^{-2} s^{-1}) and it is probably the single most important parameter to know when thinking about flow–sea bed interactions. It is commonly estimated from vertical profiles of time-averaged velocity or measurements of turbulence in the BBL (e.g. Kim et al., 2000) and is often reported as shear velocity ($u_* = \sqrt{\tau/\rho}$ where ρ is the fluid density). The reason for this conversion is that the shear stress is expressed in units of velocity and therefore can be

Box 1.2 The benthic boundary layer

A fundamental characteristic of the flow close to the seabed is the presence of a benthic boundary layer (BBL) and the flow characteristics of this layer are worth describing because this is where interactions between the sediments and water column actually happen. If we assume a constant current velocity then as we approach the seabed the frictional effects of the sediment extract momentum and the flow velocity begins to slow down until at the seabed (with stationary sediment) it approaches zero (Figure 1.2).

This reduction in flow changes the characteristics of the fluid environment and the BBL can be divided up into different sections based on these. The outer logarithmic layer is characterised by a log decline in the mean flow velocity toward the bed. Nominally the top of this layer is at 99% of the freestream flow speed and represents the extent of the

seabed influence on flow speed. In this layer, vertical mixing is rapid because of turbulence (high-frequency irregular motions—eddies—of water parcels) that pervades the fluid undergoing shear at the bed. As we move closer to the seabed, the frictional effects of the sediment begin to exert a greater impact on the fluid and the turbulence is dampened. This is called the viscous sublayer and vertical mixing in this layer is a combination of much slower molecular diffusion and smaller-scale turbulence and as a consequence is much slower. Finally, right close to the bed we have the diffusive sublayer which is where the frictional effects have dampened out all the turbulence and viscous effects (stickiness between fluid layers) dominate. This means flow tends to follow streamlines, turbulence is absent and vertical transport of solutes is dominated by molecular diffusion.

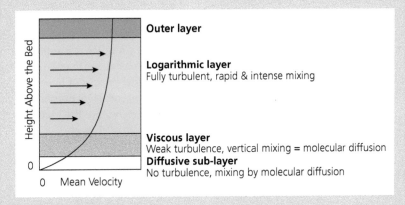

Outer layer

Logarithmic layer
Fully turbulent, rapid & intense mixing

Viscous layer
Weak turbulence, vertical mixing = molecular diffusion
Diffusive sub-layer
No turbulence, mixing by molecular diffusion

Figure 1.2 Different layers and flow characteristics of the benthic boundary layer.

used to estimate a Reynolds roughness number (Re_*) which characterises the nature of the flow at the seabed. When calculating Re_* the appropriate length scale will be the feature that dominates the seabed roughness at the scale of interest, most commonly median grain diameter but it can also be ripples or biogenic structure protruding off the seabed (for plants and fauna; see Chapter 2). When $Re_* < 3.5$ the BBL will be hydraulically smooth with viscous and diffusive sublayers, but at values >100 the flow is hydraulically rough and turbulent with no sublayers. The nice thing about Re_* (and other dimensionless numbers) is that it highlights how different combinations of bed roughness and u can generate similar BBL conditions.

A key modifier of BBL dynamics in coastal settings is the presence of short-period (1–5 s) wind-driven waves and longer-period (10–50 s) swell waves on exposed coasts. When the wave orbitals penetrate to the bed (a function of the water depth, wave height and frequency), they add a periodic oscillation of flow acceleration and deceleration in the BBL created by the mean current. These oscillations generate additional shear in the BBL and as a consequence wave–current boundary layers are virtually always hydraulically rough. The detailed physics of the wave–current boundary layer are complicated but it is worthwhile noting their importance for ecological processes. The additional shear generated by waves is often needed to initiate sediment resuspension and the associated transport/ dispersal of organic matter, sediment porewater nutrients and organisms. As the waves are primarily generated by passing weather systems this introduces a stochastic element into the frequency and intensity of sediment transport (see Hall, 1994 for an ecologically focussed review).

1.4 Consequences of the BBL on bio-physical processes

The propensity of a sediment to be mobilised is a function of the sediment grain size (and sorting), seabed roughness and the force exerted on it by the BBL. Figure 1.3 shows a simplified relationship between beds of uniform grain size and the flow velocity required to erode them. There are fundamental differences in the response of non-cohesive and cohesive sediments to increasing bed shear stress. Non-cohesive sandy sediment erosion initially starts as bedload with grains bouncing along the sediment surface (taking microbes and animals with them). As the force increases there is enough energy to move particles into suspension and then

they can be carried along with the currents. Naturally, as grain size increases the shear stress required to erode non-cohesive sediment also increases, a simple product of the fact that larger particles are heavier and more difficult to move. The behavioural response of cohesive sediments to increasing shear stress is different and they behave more like a bed of jelly than individual particles. As the erosion threshold (the point at which the applied shear stress causes the sediment bed to release particles) is exceeded, sediment is ejected (often as aggregates) directly into the water column (i.e. there is no bedload transport). Also, somewhat counter-intuitively, the force required to erode a cohesive bed *increases* with decreasing grain size because of increasing inter-particle electromagnetic forces and decreasing bed roughness which reduces the force that can be exerted on particles.

Much of our understanding of sediment resuspension and critical erosion thresholds is based on laboratory studies of single grain sizes. In the real world, coastal sediments are often comprised of mixed assemblages of fine and coarse grains which alters the cohesive properties, and the ability of water to penetrate the seabed, altering the erosion

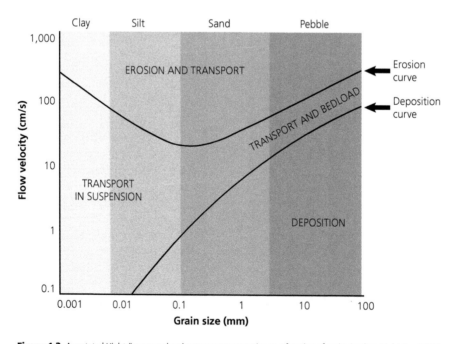

Figure 1.3 Annotated Hjulström curve showing transport properties as a function of grain size (see Hjulström, 1935).

dynamics. For example, Bartzke et al. (2013) showed that even small amounts of silt added to a sandy sediment (a few percent by weight, and well below the amount needed to cause cohesion) filled the void spaces between sand grains and substantially increased the flow speed needed to mobilise the bed. Superimposed on this is the capability of benthic organisms to alter the erosion properties of sediments (see Chapter 2). This means it is very difficult to predict erosion thresholds for coastal sediments based only on the physical grain size distribution.

At flows below the erosion threshold, the BBL can still play a crucial role in regulating solute and particle fluxes across the sediment–water interface. In permeable sediments (i.e. $K > \sim 10^{-12}$ m^2) pressure differentials generated by passing waves and/or flow over topography such as ripples induce regions of high and low pressure in the sediment, which drives a porewater exchange between the sediment (to depths of several cm) and the overlying

water column. The magnitude of this exchange is a function of the sediment permeability and the flows/topography generating the pressure gradient (Huettel & Webster, 2001), but in terms of solute exchange it is many times more efficient than molecular diffusion. Advective porewater exchange also introduces organic matter (Figure 1.4) into the sediment simultaneously with oxygen where it undergoes rapid decomposition by microbes. As a result, nutrients are released back to the water column. Santos et al. (2012) have estimated that the process of advective porewater exchange in permeable coastal sediments results in the entire volume of the ocean being filtered every 3000 years. It is this process combined with high rates of primary production which contributes to the importance of coastal sediments in global biogeochemical cycles.

The sediment surface is a region of strong and active vertical gradients in reactive solutes such as oxygen (see section 1.5). In non-permeable cohesive sediments (i.e. $K < \sim 10^{-12}$ m^2) where molecular

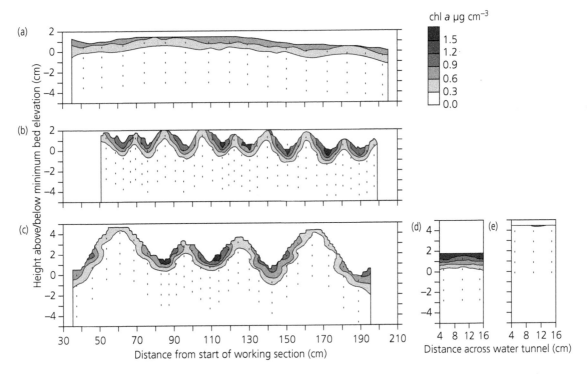

Figure 1.4 Diatom biomass measured by chlorophyll a accumulations with increasing bed geometry (a–c). (d) and (e) show a cross section through a large ripple trough and crest respectively. From Figure 1, Pilditch and Miller (2006). Reprinted from *Continental Shelf Research*, **26** (15), Pilditch C A, and Miller D C, Phytoplankton Deposition to Permeable Sediments under Oscillatory Flow: Effects of Ripple Geometry and Resuspension, 1806–1825., © 2006 reused with permission from Elsevier.

diffusion physically dominates solute exchange, BBL dynamics can still influence solute concentration gradients. The demand for oxygen in marine sediments is high due to aerobic bacterial activity and the oxidation of reduced solutes moving up from deeper in the sediments. This demand creates an oxygen gradient not only in the sediment but one that can extend into the sublayers of the BBL. The oxygen gradient in the near-bed water is known as the diffusion boundary layer (DBL) and its thickness (typically <1 mm) is regulated by a combination of the sediment oxygen demand, concentration in the overlying water column and the BBL dynamics. Because the DBL is governed in part by the sediment oxygen demand, its thickness may not correspond directly to the diffusive and viscous sublayers of the BBL. Under slow flow, this concentration gradient can extend beyond the diffusion sublayer of the BBL. Under fast, turbulent flows the DBL may disappear altogether. The presence or absence of the DBL will influence oxygen availability in surface sediments with implications for aerobic bacterial activity (see Glud, 2008 for an in-depth discussion).

Near-bed flows also play a critical role in the dispersal of benthic invertebrates and, as such, an important role in recovery from disturbances and meta-community dynamics (Chapter 3). Although dispersal is discussed in many places within the book, related to aspects of recovery, design and biotic interactions, here we briefly describe it from a flow perspective. Traditionally, the pelagic larval phase was considered most important for dispersal as this was a period that they could potentially be moved long distances by currents. However, storms may move organisms around passively and it is not uncommon to see benthic animals washed up on beaches. Under more typical conditions behaviour can influence the effect of waves and currents, e.g. crawling onto the sediment surface can increase the chances of hitching a ride with the flow (e.g. Lundquist et al., 2004). This is often particularly important for juvenile life stages where the organisms are small and easily transported. For example, the bivalve *Macomona* when it is 0.5–3 mm long will move onto the sediment surface and orientate itself so that its shell behaves a bit like a wing; once it is ready to go it excretes long proteinaceous threads

that act like parachutes lifting it off the sediment surface, allowing dispersal over large scales (100s–1000s of m). Many benthic species exhibit this post-settlement dispersal, and what is clear from field studies is that the diversity and magnitude of dispersers are functions of the near-bed hydrodynamics (e.g. Lundquist et al., 2006; Valanko et al., 2010; see Pilditch et al., 2015 for a review). There is still much to be done to understand how temporal and spatial variability in the interacting factors regulating dispersal (sediment properties, behaviour, species interactions and hydrodynamics) influence dispersal but such studies are critical to understanding the dynamics of benthic populations and communities (Pilditch et al., 2015).

1.5 Organic matter

The quality and quantity of the organic matter arriving at the sediment surface are critical environmental variables, providing fuel directly for metazoans or indirectly by fuelling microbes for decomposition that then become food for others. In general, the organic content of sediments increases with decreasing grain size because the lower flows allow the accumulation of low-density particles but also the increased surface area promotes high bacterial biomass.

In coastal environments, organic matter has numerous sources: primary production in the water column, detritus from macrophytes such as kelp, seagrass and mangroves as well as inputs from terrestrial ecosystems. In many benthic ecosystems, organic matter fuels secondary production. Except in the most turbid regions, the production of the microphytobenthos at the sediment–water interface is the major source of organic material (Cahoon, 2002; Hope et al., 2020; Miller et al., 1996).

The input of organic matter is the base of the food chain of marine sediments and the amount varies with many factors, including the productivity of the overlying water column. Food quality is a big issue and can be coarsely quantified by the carbon to nitrogen ratio. In general phytoplankton and bacteria have lower C:N ratios, typically <10, whereas macrophyte detritus typically has C:N ratios >20. Consumers prefer energy sources with lower C:N ratios (ideally as close to seven as possible, the ratio in animal tissue) because higher ratios mean more time (and energy) are expended acquiring sufficient food to satisfy their N requirements. As well as these coarse measures of food quality, it is the plants that synthesise the essential fatty acids needed by animals (Antonio & Richouz, 2014; Galloway et al., 2012). Faecal pellets from animals living in the water column or suspension feeders living on the seabed are an important source of organic matter input into seafloor communities. Faecal pellets have much higher settling velocities (e.g. Giles & Pilditch, 2004) than the food that produces them and even after digestion (and any decomposition occurring on the way down) have a higher organic content than water column particles naturally settling out. This biological pump is an important component of benthic–pelagic coupling: i.e. in this context, the delivery of water column organic matter to the seabed where it is remineralised by bacteria, releasing nutrients that fuel pelagic production (Graf & Rosenberg, 1997).

1.6 Light and benthic primary production

In shallow coastal waters where light hits the seabed the microphytobenthos and larger macrophytes (macroalgae and seagrasses) can flourish and give rise to local production. The amount of light hitting the seabed is a function of atmospheric conditions, latitude, time of day and coastal topographic features that create shade, water depth and clarity. Water clarity is strongly affected by phytoplankton in the water column, dissolved organic matter like tannins from land that stain the water and importantly the input and resuspension of fine sediments. The amount of light hitting the seafloor regulates benthic primary production and in turbid estuaries the loss of this production has been a contributing factor to accelerating eutrophication (Chapter 11).

Although on a per-area basis primary production by benthic macrophytes can greatly exceed that of unvegetated habitats, it is the ubiquitous microphytobenthos (MPB) that fuel many coastal foodwebs (e.g. Christianen et al., 2017; Jones et al., 2017). MPB consist of diatoms, dinoflagellates and cyanobacteria (McIntyre et al., 1996) growing within the first several mm of the surface sediment and can often

be seen as a greenish or brownish tinge. In more oligotrophic settings MPB production can exceed pelagic production because they have ready access to nutrients stored in the sediment porewater. Their value in coastal foodwebs arises because they are easily digestible, rich in lipids and proteins, have high turnover rates and have not undergone pre-processing by pelagic consumers and so do not arrive as faecal pellets (Hope et al., 2020). In the Dutch Wadden Sea a stable isotope study showed that MPB were the most important food source for benthic invertebrates and given these organisms support higher trophic levels (fish, birds and seals) this further underlines their importance (Christianen et al., 2017).

MPB are ecologically important beyond their value as food for higher trophic levels (Hope et al., 2020). Light is attenuated rapidly in sediments and a lack of nutrients in the photic zone means MPB undertake vertical migrations to access them (Consalvey et al., 2004). To aid these migrations, diatoms in particular excrete extracellular polysaccharides (EPS), a sticky mucus-like substance that binds the sediment together. In cohesive sediments (or non-cohesive sediments with a substantial silt/clay content) the EPS can fill the void spaces, binding sediment and particles, thus leading to a marked increase in sediment stability, making it less prone to erosion (see Chapter 2). Photosynthesis in surface sediment layers can also alter the distribution of oxygen, increasing the volume of sediment that then supports more efficient decomposition of organic matter by aerobic bacteria, affecting nutrient cycling (see section 1.8). MPB also trap nutrients at the sediment surface, preventing their release to the water column. In shallow coastal systems undergoing eutrophication, this sink of nutrients helps slow the eutrophication spiral where continual release of nutrients from decomposing organic matter in the sediments fuels further pelagic production.

1.7 Sediment biogeochemistry

Even at small scales (m²) coastal marine sediments are extremely heterogeneous with respect to their physical properties (e.g. grain size, sorting) and concentrations of reactive solutes such as oxygen.

This heterogeneity arises primarily as a result of the interactions between the activities of the resident macrofauna community and the microbial community (both the bacteria that are decomposing organic matter and the photosynthesising MPB). For example, burrow irrigation by shrimps (e.g. D'Andrea & DeWitt, 2009) may deliver oxygen deep into the sediments whereas bulldozing heart urchins leave faecal pellets that become hotspots of bacterial activity (Lohrer et al., 2004; Solan & Wigham, 2005). In Chapter 2 we explore more animal–sediment interactions but just as we have described the physical aspects of the sediment, there is value in considering the basic chemical implications of bacteria and organic matter down the sediment column.

The upper few centimeters of the sediment are one of the most extreme chemical gradients on the planet where transition occurs from an aerobic to an anaerobic environment. These gradients are driven by the physical characteristics of the sediment as well as the microbial communities and, in a world without large plants and animals, are vertically structured according to the bacterial respiration pathways during the decomposition of organic matter. The coastal sediments (which comprise <10 percent of the oceanic seafloor) are estimated to be responsible for processing ~30 percent of the world's oceanic carbon (Smith & Hollibaugh, 1998) and during the decomposition regenertaing essential nutrients. As we will see later in the book,

human-induced changes in sediment organic matter decomposition and nutrient cycling pathways can extend into the water column, negatively impacting the entire ecosystem.

If you were to carefully dig a small hole in a typical coastal sediment you would most likely see three distinct layers distinguished by their colour (and smell). These represent substantial shifts in the metabolic pathways of bacteria responsible for organic matter decomposition. The first major active zone in most coastal sediments is the surface oxic layer which appears as a yellow/brown layer that morphs into grey before immediately being replaced by reduced black anoxic sediment with a smell of rotten eggs (hydrogen sulphide). In the decomposition pathways, aerobic respiration in the surface yellow/brown layer releases the most energy and aerobes will outcompete other respiration pathways. The concentration of oxygen in a microbially dominated sediment decreases exponentially from the sediment surface to the redox potential discontinuity (RPD) where no free oxygen exists (Figure 1.5). The availability of oxygen is measured by the redox potential (Eh); once the oxygen is used up anaerobic processes dominate and control chemical speciation.

The thickness of the oxic reaction zones in sediments is variable and as a consequence so too is the efficiency with which organic matter is remineralised. The thickness of the oxic layer is a balance between supply and demand, the sediment grain

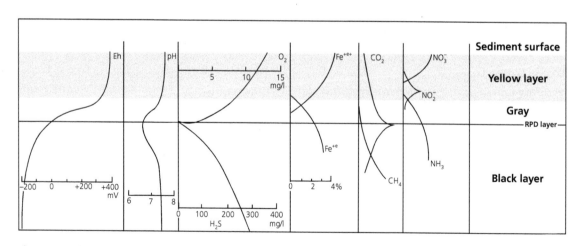

Figure 1.5 Sediment profiles of redox potential, pH, oxygen, hydrogen sulphide, iron, methane, carbon dioxide and nitrogenous species.

size which determines diffusion rates, and the quality and quantity of organic matter which determine the rates of bacterial metabolism. In finer-grained sediments diffusion rates are slower and oxygen decreases more rapidly. In permeable sediments diffusion rates are higher but there is also advective porewater exchange driven by pressure differentials. If organic loading is high, then the RPD can extend all the way to the surface and the sediments (and even the overlying water) become anoxic, limiting the higher and larger lifeforms. The loss of the oxic sediment layer also decreases the efficiency of organic matter degradation metabolism and alters the cycling of key nutrients (see section 1.8). In terms of impacts in coastal waters, increased sedimentation in many parts of the world plus eutrophication are impacting on the biogeochemistry (see Chapter 11). The presence of MPB can also induce diurnal changes in oxygen availability and during light reactions the production of oxygen can lead to a sub-surface maximum which disappears at night. Fluctuations in light intensity result in temporal variations in oxygen availability (similar fluctuations are seen through the activities of larger fauna—see Chapter 2) and may influence nitrogen cycling, a process that depends on coupled reactions in the presence and absence of oxygen (see section 1.8).

Beneath the RPD, in the absence of oxygen, other terminal electron acceptors (oxidants) are utilised by anaerobic bacteria sequentially with depth based on their standard redox potential and yield of free energy (Table 1.2; Figure 1.5). This creates a depth sequence in the oxidants consumed for the mineralisation of organic matter by bacteria, from oxygen to nitrate, manganese oxide, iron oxides, sulphate and finally carbon dioxide. The by-product of sulphate reduction is hydrogen sulphide, which gives the sediment that rotten eggs smell. Hydrogen sulphide and sulphide ions are toxic to nearly all metazoans so organisms living beneath the RPD layer obviously need to pump oxygen down. Because the oxic layer in sediments is relatively small in most of the coastal ocean it is estimated that most of the organic matter is decomposed via manganese, iron and sulphate reduction in anoxic sediments (Canfield, 1993; Jørgensen, 1982). The fate of organic matter as it passes through these different levels determines burial rates and ultimately

Table 1.2 Organic matter (CH_2O) oxidation pathways mediated by bacteria in the sediment and their free energy yields ($\Delta G°$). Table adapted from Boudreau and Jørgensen (2001)

Reaction	kJ mol^{-1}
Oxic respiration: $CH_2O + O_2 \rightarrow CO_2 + H_2O$	−479
Denitrification: $5CH_2O + 4NO_3^- \rightarrow 2N_2 + 4HCO_3^- + CO_2 + 3H_2O$	−453
Mn-oxide reduction: $CH_2O + 3CO_2 + H_2O + 2MnO_2 \rightarrow 2Mn^{2+} + 4HCO_3^-$	−349
Fe-oxide reduction: $CH_2O + 7CO_2 + 4Fe(OH)_3 \rightarrow 4Fe^{2+} + 8HCO_3^- + 3H_2O$	−114
Sulphate reduction: $2CH_2O + SO_4^{2-} \rightarrow H_2S + 2HCO_3^-$	−77
Methane production: $HCO_3^- + 4H_2 + H^+ \rightarrow CH_4 + 3H_2O$ $CH_3COO^- + H^+ \rightarrow CH_4 + CO_2$	−136 −28

the quantity of carbon stored in the sediment (Middleburg, 2019).

The by-products of anaerobic respiration/decomposition include reduced inorganic compounds in the porewater that diffuse upwards toward the sediment surface. It is interesting to note that as you move deeper in the sediment more free energy (once oxidised) is tied up in the reduced compounds, a point highlighted by the fact that methane is a by-product of the fermentation processes at depth. As these reduced compounds diffuse toward the RPD chemosynthesis can occur whereby bacteria mediate the oxidation of reduced iron, manganese and in particular sulphides to release energy that is then used to fix inorganic carbon. This form of primary production is similar to that which fuels hydrothermal vent ecosystems in the deep sea. In shallow water depths where the RPD extends to the sediment surface and light penetrates to the seabed phototrophic sulphur bacteria and cyanobacteria are also capable of carbon fixation, utilising the diffusing sulphides (see Jørgensen et al., 2019 for a recent review).

1.8 Nutrient cycling

The decomposition of organic matter within the sediments results in the regeneration of inorganic nutrients essential for primary production. It has been estimated that in some coastal and shelf ecosystems up to 30–50 per cent of the pelagic primary production is supported by nutrients regenerated from within the sediments (Nixon, 1981; Pilskaln et al., 1998). It is beyond the scope of this section to describe in detail the cycling of all biologically important elements and we refer readers to excellent summaries in Aller (1994), Middleburg (2019; carbon), Jørgensen et al. (2019; sulphur) and Delaney (1998; phosphorus) for entry points into this literature. We have chosen however to provide some detail on the sediment nitrogen (N) cycle because it is the main nutrient limiting primary production in coastal waters, of which the sediment is a key source (Nixon, 1981). Also, the sediment represents the most important biological pathway to remove excess N, which is the cause of eutrophication (see Chapter 11).

Bacteria in the sediment mediate multidirectional oxidation and reduction reactions that transform organic N into a number of different inorganic forms. A key feature regulating these transformations is the tight coupling between the oxic and anoxic zones in the sediment. The sediment porewater is rich in ammonium from the decomposition process (ammonification) and excretion by organisms. Nitrification, purely an aerobic process, transforms the ammonium into nitrate via nitrite. At each

stage of the transformation process inorganic N can diffuse upwards where it can be utilised by MPB at the sediment–water interface and, if not trapped there, by pelagic primary producers. Ammonium, nitrite and nitrate can also diffuse downwards (the direction is dependent on the concentration gradient and therefore a function of solute sources, sinks and diffusion rates) into the anoxic sediment where it can undergo a series of reduction reactions.

Of particular significance to systems experiencing excess N inputs are the pathways that reduce N to biologically inert forms. Denitrification converts nitrite and nitrate to nitrogen or nitrous oxide gas which are not readily assimilated by primary producers and are ultimately released back into the atmosphere. In the anoxic zone, dissimilatory nitrate reduction to ammonium (DNRA) can also occur where it can diffuse back into the oxic sediments or in the presence of nitrite be converted via anaerobic ammonium oxidation (anammox) to nitrogen gas (see Devol, 2015 for a review). The importance of anammox as a pathway of N removal remains relatively unknown in marine sediments, whereas in coastal sediments denitrification tends to be quantified more frequently. The small number of studies focussed on anammox in coastal sediments suggest it can account for 10–40 per cent of the total N_2 production (Devol, 2015). The removal

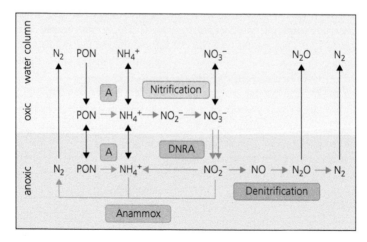

Figure 1.6 A conceptual model of the sediment nitrogen cycle in coastal sediments showing transformations specific to the oxic and anoxic sediment layers. Note that nitrification is carried out by aerobic bacteria and provides the nitrate that is converted to inert forms of N in the anoxic sediments. When the oxic layer is lost from sediments (e.g. due to organic loading) the ability of the sediments to remove bio-available N is compromised. A: ammonification; Anammox: anaerobic ammonium oxidation; DNRA: dissimilatory nitrate reduction to ammonium; PON: particulate organic nitrogen. Adapted from Devol (2015).

of nitrogen through denitrification in estuaries can account for 10–80 per cent of the inputs, thereby representing a significant sink providing resilience to eutrophication (Nixon et al., 1996).

The ability of sediments to remove excess N via denitrification is dependent on the presence of an oxic–anoxic interface. The simple one-dimensional picture of N-cycling in coastal sediments shown in Figure 1.6 masks the complexity introduced by the activities of larger fauna that can greatly extend this interface through building and irrigating burrows or bulldozing their way through sediment, both activities that can stimulate denitrification (see Chapters 2 and 9). Often in eutrophic systems excess organic matter inputs result in the loss of the sediment oxic layer (RPD rises into the water column). Under these conditions denitrification is halted and the sediments become a significant source of regenerated ammonium that continues to accelerate primary production.

1.9 Close out

A lot of processes are interacting in marine sediments and the overlying water. We have highlighted the intermittent interplay between the physical processes of water flow, sediment grain size and the supply and decomposition of organic matter. These processes regulate the chemistry of this reactive layer and make a significant contribution to global cycles of nutrients and carbon, despite the biologically active layer being only a few centimeters deep. In coastal waters, the pelagic and benthic environments are tightly coupled and solute and particle exchanges across the sediment–water interface regulate important functions. Thus, the benthic and pelagic realms need to be seen as one system. Conversely, in deep-sea habitats the benthic influence on pelagic habitats is lost (but not vice versa) because of water circulation times.

The physico-chemical processes we have highlighted are often well supported by theory but we have limited empirical data. This can result in us missing important processes (e.g. anammox was only discovered about 20 years ago; Kuenen, 2008). Our understanding of how these basic processes operate can also be biased by data being collected in a restricted geographic region (Vieillard et al., 2020).

Therefore, although these processes have been under study for many years, there is still plenty of work to be done. However, physico-chemical processes do not operate in a vacuum. Although there are 'dead' zones in our marine systems, they are fortunately rare and instead physico-chemical processes interact with, and are often driven by, the plants and animals that live in and on the sediment (see Chapter 2). The interplay between these factors is an area rich in opportunities for future research.

References

Aller R C (1994). Bioturbation and Remineralization of Sedimentary Organic Matter: Effects of Redox Oscillation. *Chemical Geology*, **114**, 331–45.

Antonio E and Richoux N (2014). Trophodynamics of Three Decapod Crustaceans in a Temperate Estuary using Stable Isotope and Fatty Acid Analyses. *Marine Ecology Progress Series*, **504**, 193–205.

Bartzke G, Bryan K R, Pilditch C A and Huhn K (2013). On the Stabilizing Influence of Silt on Sand Beds. *Journal of Sedimentary Research*, **83**(8), 691–703.

Boudreau B P and Jørgensen B B, eds. (2001). *The Benthic Boundary Layer: Transport Processes and Biogeochemistry*, Oxford University Press, Oxford.

Cahoon L B (2002). The Role of Benthic Microalgae in Neritic Ecosystems. *Oceanography and Marine Biology*, **37**, 47–86.

Canfield D E (1993). Organic Matter Oxidation in Marine Sediments. In: Wollast R, Mackenzie F T and Chou L, eds. *Interactions of C, N, P, and S Biogeochemical Cycles and Global Change*, Springer-Verlag, Berlin, pp. 333–63.

Christianen M J A, Middelburg J J, Holthuijsen S J, Jouta J, Compton T J, van der Heide T, Piersma T, Sinninghe Damsté J S, van der Veer H W, Schouten S and Olff H (2017). Benthic Primary Producers are Key to Sustain the Wadden Sea Food Web: Stable Carbon Isotope Analysis at Landscape Scale. *Ecology*, **98** (6), 1498–512.

Consalvey M, Paterson D M and Underwood G J C (2004). The Ups and Downs of Life in a Benthic Biofilm: Migration of Benthic Diatoms. *Diatom Research*, **19** (2), 181–202.

D'Andrea A F and DeWitt T H (2009). Geochemical Ecosystem Engineering by the Mud Shrimp Upogebia pugettensis (Crustacea: Thalassinidae) in Yaquina Bay, Oregon: Density-Dependent Effects on Organic Matter Remineralization and Nutrient Cycling. *Limnology and Oceanography*, **54**, 1911–32.

Delaney M L (1998). Phosphorus Accumulation in Marine Sediments and the Oceanic Phosphorus Cycle. *Global Biogeochemical Cycles*, **12**(4), 563–72.

Devol A H (2015). Denitrification, Anammox, and N$_2$ Production in Marine Sediments. *Annual Review of Marine Science*, **7**, 403–23.

Douglas E J, Pilditch C A, Lohrer A M, Savage C, Schipper L A and Thrush S F (2018). Sedimentary Environment Influences Ecosystem Response to Nutrient Enrichment. *Estuaries and Coasts*, **41**(7), 1994–2008.

Folk R L (1954). The Distinction between Grain Size and Mineral Composition in Sedimentary-Rock Nomenclature. *The Journal of Geology*, **62**(4), 344–59.

Galloway A W E, Britton-Simmons K H, Duggins D O, Gabrielson P W and Brett M T (2012). Fatty Acid Signatures Differentiate Marine Macrophytes at Ordinal and Family Ranks. *Journal of Phycology*, **48**(4), 956–65.

Giles H and Pilditch C (2004). Effects of Diet on Sinking Rates and Erosion Thresholds of Mussel *Perna canaliculus* Biodeposits. *Marine Ecology Progress Series*, **282**, 205–19.

Glud R N (2008). Oxygen Dynamics of Marine Sediments. *Marine Biology Research*, **4**(4), 243–89.

Graf G and Rosenberg R (1997). Bioresuspension and Biodeposition: A Review. *Journal of Marine Systems*, **11**(3), 269–78.

Green M O and Coco G (2014). Review of Wave-Driven Sediment Resuspension and Transport in Estuaries. *Reviews of Geophysics*, **52**(1), 77–117.

Hall S J (1994). Physical Disturbance and Marine Benthic Communities: Life in Unconsolidated Sediments. *Annual Review of Oceanography and Marine Biology*, **32**, 179–239.

Hjulström F (1935). Studies of the Morphological Activity of Rivers as Illustrated by the River Fyris. *Bulletin of the Geological Institute of Uppsala*, **25**, 221–527.

Hope J A, Paterson D M and Thrush S F (2020). The Role of Microphytobenthos in Soft-Sediment Ecological Networks and Their Contribution to the Delivery of Multiple Ecosystem Services. *Journal of Ecology*, **108**(3), 815–30.

Huettel M and Webster I T (2001). Porewater Flow in Permeable Sediment. In: Boudreau B P and Jørgensen B B, eds. *The Benthic Boundary Layer: Transport Processes and Biogeochemistry*, Oxford University Press, Oxford, pp. 144–79.

Jacobs W, Le Hir P, Van Kesteren W and Cann P (2011). Erosion Threshold of Sand–Mud Mixtures. *Continental Shelf Research*, **31** (10, Supplement), S14–25.

Jones H F E, Pilditch C A, Hamilton D P and Bryan K R (2017). Impacts of a bivalve Mass Mortality Event on an Estuarine Food Web and Bivalve Grazing Pressure. *New Zealand Journal of Marine and Freshwater Research*, **51**(3), 370–92.

Jørgensen B B (1982). Mineralization of Organic Matter in the Sea Bed—the Role of Sulphate Reduction. *Nature*, **296** (5858), 643–5.

Jørgensen B B, Findlay A J and Pellerin A (2019). The Biogeochemical Sulfur Cycle of Marine Sediments. *Frontiers in Microbiology*, **10**, 849.

Jumars P A (1993). *Concepts in Biological Oceanography: An Interdisciplinary Primer*, Oxford University Press, New York.

Kamann P J, Ritzi R W, Dominic D F and Conrad C M (2007). Porosity and Permeability in Sediment Mixtures. *Groundwater*, **45**(4), 429–38.

Kenny A J and Sotheran I (2013). Characterising the Physical Properties of Seabed Habitats. In: Eleftheriou A, ed. *Methods for the Study of Marine Benthos*, 4th ed., John Wiley & Sons, Hoboken, NJ, pp. 47–95.

Kim S C, Friedrichs C T, Maa J P Y and Wright L D (2000). Estimating Bottom Stress in Tidal Boundary Layer from Acoustic Doppler Velocimeter Data. *Journal of Hydraulic Engineering*, **126**(6), 399–406.

Klute A and Dirksen C (1986). Hydraulic Conductivity and Diffusivity. Laboratory Methods. In: Klute A, ed. *Methods of Soil Analysis—Part 1. Physical and Mineralogical Methods*, American Society of Agronomy, Madison, WI, pp. 687–734.

Kuenen J G (2008). Anammox Bacteria: From Discovery to Application. *Nature Reviews Microbiology*, **6** (4), 320–6.

Le Hir P, Monbet Y and Orvain F (2007). Sediment Erodability in Sediment Transport Modelling: Can We Account for Biota Effects? *Continental Shelf Research*, **27**(8), 1116–42.

Leeder M R (1982). *Sedimentology: Process and Product*, G. Allen & Unwin, London.

Lohrer A M, Thrush S F and Gibbs M M (2004). Bioturbators Enhance Ecosystem Function through Complex Biogeochemical Interactions. *Nature*, **431** (7012), 1092–5.

Long D (2006). *BGS Detailed Explanation of Seabed Sediment Modified Folk Classification*.

Lundquist C J, Pilditch C A and Cummings V J. (2004). Behaviour Controls Post-Settlement Dispersal by the Juvenile Bivalves *Austrovenus stutchburyi* and *Macomona liliana*. *Journal of Experimental Marine Biology and Ecology*, **306**(1), 51–74.

Lundquist C J, Thrush S F, Hewitt J, Halliday J, MacDonald I and Cummings V (2006). Spatial Variability in Recolonisation Potential: Influence of Organism Behaviour and Hydrodynamics on the Distribution of Macrofaunal Colonists. *Marine Ecology Progress Series*, **324**, 67–81.

MacIntyre H L, Geider R J and Miller D C (1996). Microphytobenthos: The Ecological Role of the "Secret Garden" of Unvegetated, Shallow-Water Marine Habitats. I. Distribution, Abundance and Primary Production. *Estuaries*, **19**(2), 186–201.

Middleburg J J (2019). *Marine Carbon Biogeochemistry: A Primer for Earth System Scientists*, Springer International, Cham.

Miller D C, Geider R J and MacIntyre H L (1996). Microphytobenthos: The Ecological Role of the "Secret Garden" of Unvegetated, Shallow-Water Marine Habitats. II. Role in Sediment Stability and Shallow-Water Food Webs. *Estuaries*, **19**(2), 202–12.

Nixon S W (1981). Remineralization and Nutrient Cycling in Coastal Marine Exosystems. In: Neilson B J and Cronin L E, eds. *Estuaries and Nutrients*, Humana Press, Totowa, NJ, pp. 111–38.

Nixon S W, Ammerman J W, Atkinson L P, Berounsky V M, Billen G, Boicourt W C, Boynton W R, Church T M, Ditoro D M, Elmgren R, Garber J H, Giblin A E, Jahnke R A, Owens N J P, Pilson M E Q and Seitzinger S P (1996). The Fate of Nitrogen and Phosphorus at the Land-Sea Margin of the North Atlantic Ocean. *Biogeochemistry*, **35** (1), 141–80.

Nowell A R M and Jumars P A (1984). Flow Environments of Aquatic Benthos. *Annual Review of Ecology and Systematics*, **15**, 303–28.

Pilditch C A and Miller D C (2006). Phytoplankton Deposition to Permeable Sediments under Oscillatory Flow: Effects of Ripple Geometry and Resuspension. *Continental Shelf Research*, **26** (15), 1806–25.

Pilditch C A, Valanko S, Norkko J and Norkko A (2015). Post-Settlement Dispersal: The Neglected Link in Maintenance of Soft-Sediment Biodiversity. *Biology Letters*, **11**(2), 20140795.

Pilskaln C H, Churchill J H and Mayer L M (1998). Resuspension of Sediment by Bottom Trawling in the Gulf of Maine and Potential Geochemical Consequences. *Conservation Biology*, **12**(6), 1223–9.

Pratt D R, Lohrer A M, Pilditch C A and Thrush S F (2014). Changes in Ecosystem Function across Sedimentary Gradients in Estuaries. *Ecosystems*, **17**(1), 182–94.

Sanders H L (1958). Benthic Studies in Buzzards Bay. I. Animal-Sediment Relationships. *Limnology and Oceanography*, **3**(3), 245–58.

Sanders H L (1960). Benthic Studies in Buzzards Bay III. The Structure of the Soft-Bottom Community. *Limnology and Oceanography*, **5**(2), 138–53.

Santos I R, Eyre B D and Huettel M (2012). The Driving Forces of Porewater and Groundwater Flow in Permeable Coastal Sediments: A Review. *Estuarine, Coastal and Shelf Science*, **98**, 1–15.

Smith S V and Hollibaugh J T (1998). Carbon-Nitrogen-Phosphorus Cycling in Tomales Bay, California. *Aquatic Geochemistry*, **4**(3), 395–402.

Snelgrove P V R and Butman C A (1995). Animal-Sediment Relationships Revisited: Cause versus Effect. *Oceanographic Literature Review*, **42**(8), 668.

Solan M and Wigham B D (2005). Biogenic Particle Reworking and Bacterial–Invertebrate Interactions in Marine Sediments. In Kristensen E, Haese R R and Kostka J E, eds. *Interactions between Macro- and Microorganisms in Marine Sediments*, vol. 60, American Geophysical Union, Washington, DC, pp. 105–24.

Staudt F, Mullarney J C, Pilditch C A and Huhn K (2017). The Role of Grain-Size Ratio in the Mobility of Mixed Granular Beds. *Geomorphology*, **278**, 314–28.

Traykovski P, Hay A E, Irish J D and Lynch J F (1999). Geometry, Migration, and Evolution of Wave Orbital Ripples at LEO-15. *Journal of Geophysical Research: Oceans*, **104** (C1), 1505–24.

Valanko S, Norkko A and Norkko J (2010). Strategies of Post-Larval Dispersal in Non-Tidal Soft-Sediment Communities. *Journal of Experimental Marine Biology and Ecology*, **384**(1), 51–60.

Vieillard A M, Newell S E and Thrush S F (2020). Recovering from Bias: A Call for Further Study of Underrepresented Tropical and Low-Nutrient Estuaries. *Journal of Geophysical Research: Biogeosciences*, **125**(7), e2020JG005766.

CHAPTER 2

Benthic animals and plants and what they do to sediments

2.1 Introduction

In the previous chapter we presented an introduction to the sedimentary environment focussing primarily on physical characteristics and how the activity of microbial organisms alters sediment chemistry. However, any cursory look at the sediment surface will show traces of larger life forms in the sediment that underlie a much greater complexity. This can include tubes, burrows, mounds and pits created by the infauna as well as the epifaunal animals and large plants living on the sediment surface. These surface features indicate the presence of a complex three-dimensional structure extending above and below the sediment surface that often defines seafloor habitats and how they function.

Above the surface the spatial structure of emergent habitats interacts with the benthic boundary layer and in combination with the feeding activities of the animals dramatically alters benthic–pelagic coupling (i.e. the flux of dissolved nutrients/solutes, inorganic/organic particles and organisms). For example, in sufficient densities these structures can stabilise the sediment and promote the settlement of organic matter. In areas devoid of emergent biogenic structure, the feeding, movement and irrigation activities of large infaunal organisms can radically alter the sediment properties such as grain size distribution and the availability of oxygen, creating a 3-D mosaic of microhabitats that affects biogeochemistry and rates of organic matter decomposition and nutrient cycling. Sometimes single species are important and these are usually the larger species. The role of these large species had been known for a long time but the coining of the term ecosystem engineer has more widely highlighted their significance (Jones et al., 1994).

So, in this chapter we move from the simple microbial, geochemical world view presented in Chapter 1 to consider the animals and plants that live on and in the sediment (Figure 2.1). We introduce these and the general categories in which they are often studied. As these organisms frequently alter the physical environment that surrounds them, in terms of sediment characteristics and sedimentary processes and also alterations in benthic–pelagic coupling, we also discuss some essential characteristics that determine what, how and how much they alter their environment.

2.2 Size

Size is particularly important for studies of benthic organisms, in part because sampling strategies are dependent on size. For this reason, the communities that live in marine sediments are divided into different groups based on size, and these groups are often studied separately.

The microbial communities are composed of unicellular organisms (e.g. protists, archaea, diatoms). They are found throughout the sediment column and are especially important in driving the transformation of organic matter in the sediment and the processes of remineralisation (see Chapter 1). These organisms can form colonies or mats that create large structures, for example stromatolites and biofilms or mats consisting of microphytes or under reduced conditions Beggiatoa, or they can function

Ecology of Coastal Marine Sediments: Form, Function, and Change in the Anthropocene. Simon F. Thrush, Judi E. Hewitt, Conrad A. Pilditch and Alf Norkko, Oxford University Press (2021). © Simon F. Thrush, Judi Hewitt, Conrad Pilditch, and Alf Norkko.
DOI: 10.1093/oso/9780198804765.003.0002

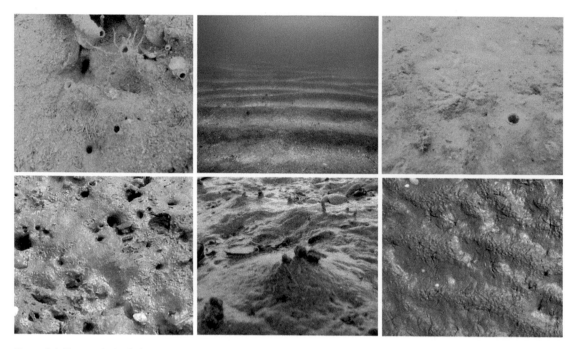

Figure 2.1 Plants and animals change the topography of the sediment surface, providing clues to burrows, tubes and processes beneath the sediment surface. Tubes, burrows, faecal mounds and feeding traces are linked to sediment type and the action of physical forces on the seafloor. (Photo credits ST, Rod Budd and Jenny Hillman.)

as individual cells. The composition and activity of microbial communities are influenced by their environmental conditions and in particular whether the sediment is oxygenated or anaerobic.

For the larger multicellular fauna, size classes are conventionally defined by the mesh size of sieves on which the animals would be retained. Meiofauna are the smallest; they pass through a 0.5-mm mesh sieve and are retained on a mesh larger than 0.062 or 0.031 mm. These are not animals you can see with the naked eye, and they usually live in the interstices between grains of sediment. They graze on detritus, microbes and recently settled macrofauna and can contribute to sedimentary processes and functions (Bonaglia et al., 2014; Coull, 1999). Nematodes, harpacticoid copepods, turbellarians and gastrotrichs are commonly represented in meiofaunal communities, but there are representatives of almost all marine invertebrate phyla within the meiofauna.

The macrofauna are classified as organisms larger than 0.5 mm, although some researchers use 0.3 mm, whereas others use 1.0 mm. This distinction is usually based on how important the juvenile vs adult life stages are to the individual researcher. For instance, if you are interested in recruitment and early benthic life stages that settle at 0.2–0.3 mm, or if you are sampling a community dominated by small species you would select a smaller mesh size; however, if your interests are in the adult life stages of larger organisms, or you want to sieve samples more quickly, a coarser mesh would be used. The early life stages of some of the larger macrofaunal species are classified as meiofauna until they grow to larger than 0.5 mm. Most kinds of marine invertebrates are represented in the macrofauna, but polychaetes, bivalves, gastropods, amphipods, isopods and ophiuroids are among the most common. These are the organisms that are big enough to start moving sediment particles and pumping water into and out of the sediment, so that they have a strong influence on ecosystem function and the sedimentary environment. There is no common definition of the upper size limit to macrofauna, but the practicalities of sampling come into play. Macrofauna are usually collected by core or grab samples and therefore are commonly less than 10 cm.

Larger fauna are called megafauna; these are often animals like fish, large crustaceans and octopuses that burrow in sediments or the large predators which feed in and disturb the sediments—which can include very large animals like grey whales and walrus. However, the cut-off point of 10 cm means that many polychaetes, ophiuroids, echinoderms and even some bivalves could be considered megafauna in their adult stages. For this reason, it is important not to take these categories too strictly.

Apart from microalgae, a variety of plants inhabit the soft-sediment seafloor; these range from macroalgae (seaweeds) to seagrasses and vascular plants. Although there are some species of red and green macroalgae adapted to grow in relatively sheltered soft sediments, they do not reach the biomass and forest-like structure associated with rocky substrate kelp forests. For example, the temperate red alga *Ademsiella chauvinii* form patches of several m² in extent and extend a few tens of cm off the bed (Kregting et al., 2008). Seagrasses and vascular plants generally extend farther off the seafloor, but this is strongly species-dependent.

Still, these different size classifications of benthic organisms are often studied by different research groups with different objectives, approaches and tools. To some extent this emphasises the practicalities of sampling different kinds of communities (Gray, 1981). But these different size spectra do exist and fundamentally represent organisms with different functions and different turnover rates that operate on different space and time scales.

No matter what your study objectives and available sampling techniques, it is important to remember that all these different size-based classifications of benthic communities interact. For example, the microphytobenthos forms part of the microbial community, but is an important food resource for meiofauna and macrofauna that in turn are eaten by fish and shore birds.

2.3 Living position

Three categories that we need to mention for completeness are related to living position: epibenthos, living on the seafloor; infauna, living within the sediment; and fringing vegetation, living in high upper tidal zones.

2.3.1 Epibenthos

Epibenthos live on the sediment surface, and are comprised of both plants and animals. Epiflora (plants) are comprised of macro-algae and seagrasses. They range from individuals to extensive beds and from encrusting to erect forms. They are generally anchored (at least in the beginning) to small hard items, such as rocks, cobbles or shell. An exception to anchoring is the rhodoliths. These are red algae but they build calcium carbonate deposits (much like the unrelated encrusting corallines). Also known as living stones, rhodoliths roll across the seafloor until they reach a size where they can no longer be transported by the prevailing waves and currents.

Seagrasses are the only marine angiosperm (terrestrial plant) to invade the marine habitats and have been the subject of an extensive array of studies. Seagrass fronds can trap passing fine sediment particles, altering the sediment type beneath the bed (De Boer, 2007; Hendriks et al., 2010). The primary production from seagrass often enters the benthic foodwebs via detritus. Beneath the sediment surface the uptake of nutrients and release of organic compounds influence microbial community structure and sediment biogeochemistry.

Epifauna consist of reefs of shellfish that sit on the sediment such as mussels and oysters, and other sedentary species such as sponges, bryozoans, soft corals and sea pens. Epifauna can also be mobile species such as small amphipods that scoot around the sediment surface or larger crustaceans, gastropods and echinoderms that roam across the sediment surface. Their major contributions to modifying the physical environment will be covered in the feeding, habitat formation and mobility sections below (sections 2.4–2.6). Regardless of whether the epibenthos are plants or animals or a mix, in soft sediments they can create an oasis of stability and provide an important transition between rocky reef and soft-sediment habitats.

2.3.2 Infauna

Infauna are meio-, macro- or mega-fauna that live within the sediment. They are represented by most of the taxonomic groups and range from individuals that are attached to small, hard substrates buried in the sediment, to species that move only rarely or over very small spatial scales on a daily basis, to species that freely burrow through the sediment. Some species will build extensive galleries within the sediment and pump water and detritus, microalgae and organic matter through the galleries. Some large shellfish will anchor themselves within the sediment by threads and protrude above the sediment surface. Other species create tubes that stick up through the surface and feed and move only within these tubes. Still others move freely either within the sediment surface, or between the deep (sometimes anoxic) and the surface, transporting sediment, oxygen and pore water. As you can see, the categorising of species into 'infauna' vs 'epifauna' can be difficult.

2.3.3 Fringing vegetation

Fringing vegetation includes mangroves (tropical and subtropical) and salt marshes (temperate). Binding and trapping of sediment by these plants has an important role in the geomorphological evolution of shorelines and providing resilience to climate change. On shorter time scales, areas covered by mangroves and salt marshes are productive with the plants providing high amounts of organic matter albeit of low nutritional (high C:N ratio) value. The leaf litter often ends up being incorporated into coastal foodwebs. In highly oligotrophic (mainly tropical) systems mangrove production is an important contributor to secondary production (often via bacterially mediated decomposition pathways).

2.4 Feeding modes of benthic animals

Benthic fauna can be classified by their feeding mode. How these animals feed drives a lot of what they do to modify the physical environment and underpins much of the functionality of marine sediments (see Chapters 9 and 10). It is often the feeding activities at an individual level that, when scaled up, are responsible for shifts in function. Broadly these animals can be divided into three important categories: predators/scavengers; suspension and filter feeders; and deposit feeders/grazers. While this is a very simple classification, although with many nuances for different species, it is nevertheless a useful one. However, it is important to remember that, in some taxa, feeding mode is plastic and able to be switched depending on environmental conditions. Regardless of the feeding mode, the dilution of organic matter by indigestible sediment poses some unique challenges to energy acquisition for benthic animals.

2.4.1 Suspension feeders

Suspension feeding is a feeding mode that has evolved in multiple benthic phyla including polychaetes, anemones, corals, bryozoans, echinoderms, crustaceans, hydroids and molluscs. Many suspension feeders live on the sediment surface (epifauna) but some also live buried in the sediment and pump seawater down to their feeding structures via tubes or siphons. The suspension feeders are broadly classified as active or passive. In passive suspension feeders, feeding appendages are exposed to the currents and particles travelling in the water column are trapped and transported to the mouth (e.g. corals, foraminifers and ophiuroids). Many of the polychaete tube worms that form dense mats are also passive suspension feeders. The energy requirements of these passive suspension feeders are low because they do not generate active currents and so can persist in regions of low or sporadic food resources.

Active suspension feeders are those that expend energy to generate water flows across structures that capture particles or engage in other feeding-related behaviours when they sense the presence of nutritious particles (e.g. ciliates, sponges, crustaceans and bivalves). Some active suspension feeders may be referred to as 'filter feeders' because they pump water through a structure that functions as a filter (e.g. sea squirts); however, benthic ecologists do not generally make a distinction between the precise 'filter feeder' and the more general 'suspension feeder' term.

The feeding ecology of suspension-feeding bivalves is particularly well studied. Oysters, mussels, clams, scallops and many other bivalves are culturally important food resources for humans. These energetic feeders live in regions of abundant food resources and at least historically they often dominated harbours and estuaries. These active suspension feeders often form distinct and dense beds, they create habitat for other organisms (see Chapter 7) and their feeding activity can be an important regulator of both benthic and pelagic systems, as demonstrated below.

Suspension-feeding bivalves use cilia on their gill structures to generate water currents that draw water in. This process can remove particles sized from several μm to hundreds of μm and depending on size achieve filtration rates of several litres per hour. When densities reach several thousand per m² this represents an important biological filter on water clarity and removal of water column productivity. The water flow on the gill surface acts not like a filter mesh taking particles out; rather, particles are bound in mucus and transported toward the mouth. Prior to ingestion particles are sorted on the labial palp and those not selected for ingestion are rejected through the exhalent current as pseudofaeces. This sorting process provides a way to reject non-nutritional particles prior to ingestion and can increase the quality of the ingested material by up to a factor of 15 (Kiorboe & Mohlenberg, 1981). But at the same time, it also removes suspended sediment from the water column to the sediment surface, bound in a protein-rich mucus. The ability to sort and reject particles is species-dependent and also varies within species dependent on particle quality, type and size (Safi et al., 2007). The gill is a sensitive organ and the ability of suspension feeders to deal with suspended sediments is a critical factor in their energetics. If the turbidity is too high gill structures can become damaged and/or simply food is diluted to a point where they cannot acquire enough energy for growth and reproduction, so they die. There are several good summaries of bivalve feeding that readers should consult (Bayne et al., 1993; Jørgensen, 1996; Rosa et al., 2018).

Many suspension feeders display strong preferences for certain particles according to size, shape or chemical properties. Moreover, although we generally think of suspension feeders as feeding on phytoplankton, in reality particles fed on can include zooplankton, larvae, bacteria, detritus and suspended microphytobenthos. Thus, some suspension feeders are carnivores (including some sponges), and others that feed at the sediment–water interface are primarily detritivores. Sponges may be the most diverse feeders of the suspension feeding group, ranging from feeding on phytoplankton to being carnivores, to using spicules to garden diatoms (Bavestrello et al., 2000), to directly consuming dissolved organic carbon (Wooster et al., 2019; Yahel et al., 2003).

An important consequence of the suspension-feeding processes is the particle transformations these organisms undertake. Faeces and pseudofaeces can change the flux of organic matter to the seabed, these particles have a lower C:N ratio than the bulk sediment. Where dense aggregations exist this added flux alters microbial activity and community structure (Norkko et al., 2001), and, interestingly, suspension feeders are more likely to form aggregations than deposit feeders (Herman et al., 1999).

2.4.2 Deposit feeders

Deposit feeding is generally the most abundant feeding type in soft-sediment systems (Lopez et al., 1989). Deposit feeders include holothurians, echinoids, gastropods, bivalves, ophiuroids, decapods and polychaetes. They were so called because it was thought that they fed on organic matter that had drifted down through the water and settled (been deposited) on the bottom. However, similar to suspension feeders, deposit feeders feed on many different types of particles originating from many different places. The one thing they have in common is that they are feeding either on the sediment surface or in the sediment.

Many of them are bulk processors, that is, they ingest sediment, absorb what organic matter they can and process the rest as faeces. Others are specialist grazers on microphytobenthos and bacteria. They use a variety of methods to collect the sediment and pull it into their mouths, including tentacles, arms or nets, and digest and assimilate a fraction of the microbial community living on those particles. Deposit-feeding bivalves generally use siphons, often with the inhalant and exhalent siphons separated, and the length of the inhalant siphon controls the depth in the sediment to which the animal can live. Tentacles, arms and siphons can be used to stir the sediment, helping separate microalgae and detritus from the sediment surface.

Deposit feeders also employ a variety of strategies in where they feed. Conveyor belt deposit feeders, or head-down bottom-up, eat at depth and defaecate at the surface. Subductive feeders, or head-up bottom-down, eat on the sediment surface and defecate at depth. However, there are also deposit feeders that eat and defecate within the sediment; these are more likely to be bulk processors. These are all important for particle transport within the sediment and aeration of the sediment (see section 2.5.1). Deposit feeders can also directly alter the sediment grain size, either by palletisation (larger particles produced by excretion) or by particle size selection during feeding. For example, many maldanid worms are conveyor belt deposit feeders. As they excavate at the bottom of their tubes, fine sand from the sediment surface falls into this excavation, changing the grain size of the sediment surface (Levin et al., 1997).

Importantly, many deposit feeders, with the exception of bulk processors, can switch to suspension feeding. This switch is usually driven by location of resources and modified by current. More generally, all these fauna respire, releasing CO_2, and excrete, increasing NH_4. They influence biogeochemistry, affecting the carbon and nitrogen supply at different depths in the sediment column (Middelburg, 2017).

2.4.3 Predators and scavengers

Predators/scavengers is the other main feeding mode in soft sediments. They are generally considered together, both because many species do both, but also because it is difficult to know which they do. Rough categorisation is usually based on jaw structures (Fauchauld & Jumars, 1979). Multiple trophic levels are created by predation and scavengers, generally related to both size of the predator and more specifically the size of its jaws and the size of prey it can consume. Although it is obvious that this group does modify the sedimentary

environment by respiration and excretion, it also provides some organic matter for further breakdown by microbes as not all of a prey item is consumed. Furthermore, excavation of, or burrowing through, the sediment by predators seeking their prey creates sediment disturbance and turnover. Localised feeding pits, often formed by crustaceans, birds or fish, which expose other infauna to predation, can range from mm to m and cover large areas. The sediment disturbance and rate of turnover (bioturbation) is controlled by size and mobility and its effect on the biogeochemistry of the sediment is discussed in section 2.5.1.

2.5 Mobility

The seafloor is a busy place. Many of the fauna are mobile, scurrying over or ploughing through the sediment surface, digging burrows or burrowing through sediment. Mobile organisms moving around, in or on the sediment, have multiple effects on both sediment characteristics and sedimentary processes either directly (bioturbation) or indirectly (sediment destabilisation).

2.5.1 Bioturbation

Since fauna first developed in the Cambrian era, the seafloor has changed from the system described in Chapter 1 to a bioturbated environment, with on average the upper 5–10 cm being biologically mixed (Boudreau, 1998). Studies of sediment reworking by animals have a long history, moving from studies of rates of particle transport by specific species (Cadee, 1979; Rhoads, 1963, 1974) to understanding the effects of this particle transport on the surrounding environment (Aller, 1982; Lohrer et al., 2004). Sediment reworking occurs through a number of activities: burrow construction and maintenance; ingestion and excretion, especially of bulk processor deposit feeders; and through larger epifauna ploughing through the sediment. This reworking results in particles moving vertically or laterally, burying or exposing organic matter and microbial organisms (Figure 2.2).

However, animals do not just rework sediments, they can also affect porewater flow (advection or bio-irrigation) through feeding activities.

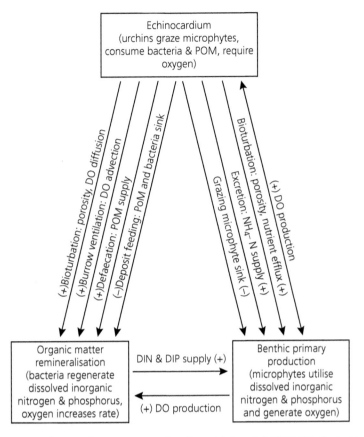

Figure 2.2 Effect of a burrowing urchin on sedimentary processes. From Figure 1, Lohrer et al. (2004). DIN: dissolved inorganic nitrogen; DIP: dissolved inorganic phosphorus; DO: dissolved oxygen; POM: particulate organic matter.

Crustaceans or polychaetes often pump water through their burrows (e.g. Aller, 1988; Kristensen, 2001) which, particularly if the burrows are closed, or burrows are permeable, generate pressure waves through the sediment (Wethey & Woodin, 2005; Volkenborn et al., 2010). Similar porewater movement generated by pressure waves has been observed for bivalve feeding with excurrent siphons below the sediment surface (e.g. tellinid bivalves: Reise, 1983; Woodin et al., 2016; see Figure 2.3). This porewater bioadvection, driven by animal activity, moves anoxic and oxic water through the sediment and varies with time in both rate and direction (Krueger, 1964). Changes in direction can lead to redox oscillations on the scale of minutes to hours (Volkenborn et al., 2010), with important consequences for sediment biogeochemistry (Aller, 1994; Timmermann et al., 2006). Volkenborn et al.

(2010) found that the area influenced by an arenicolid polychaete experienced transient bursts of activity, bio-hydraulic sediment cracking, forward and reversed advective pulses, and long-term flow reversals.

However, many animals that contribute to particle reworking also contribute to porewater advection—often by the same activities. For this reason Kristensen et al. (2012) suggested that the term bioturbation should be standardised as covering 'all transport processes and their physical effects on the substratum'. The study of bioturbation and its effects on the physical and chemical composition of sediment has been covered in many books (Aller, 1982; Kristensen, 2000; Rhoads & Boyer, 1982) and will be covered in more detail in Chapter 9. However, generally, bioturbation has been demonstrated to affect the flux of nitrogen and oxygen,

(a)   *Macomona liliana*

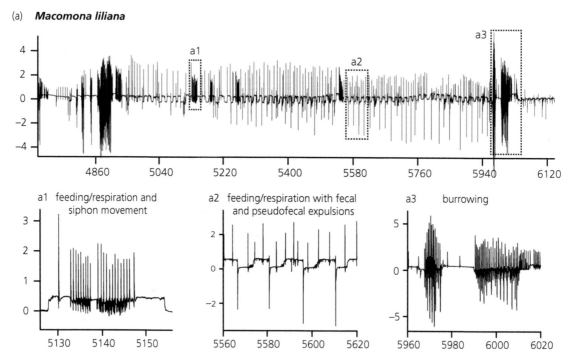

Figure 2.3 Different animal behaviours generate different pressure signals in the porewater. From Figure 2 in Woodin et al. (2016). Reprinted under Creative Commons Attribution 4.0 International (CC BY 4.0) licence.

improving conditions for production by microphytobenthos, and changing surficial sediment characteristics.

It is also important to remember that not all particle reworking or porewater pressurisation is due to animal behaviour. As described in Chapter 1, waves and currents are both drivers of these processes and the relative importance of physical and biotic drivers is a fascinating area of study (Grant, 1983). Regardless, except in extremes of these physical drivers, modification by benthic plants and animals is, at the least, of equal importance and frequently more important.

2.5.2 Sediment destabilisation

Another effect of mobility on the soft-sediment seafloor is sediment destabilisation (Needham et al., 2012; Rhoads & Young, 1970; Volkenborn et al., 2009). Animals scurrying across or bulldozing through the sediment disturb microphytes which

may otherwise stabilise the sediment and allow both sediments and microphytes to be picked up and moved by waves and currents. Animals digging and clearing burrows both lift sediment into the water, allowing it to be carried away, or create mounds on the surface that can be washed away. Different-sized particles are eroded and transported differentially, resulting in changes to the sediment particle size and organic content of the sediment.

2.6 Habitats formed by organism–environment interactions

Habitats in soft sediments are frequently described by the physical environment. Despite this, the majority of habitats in soft sediments are created by species living in or on the sediment. These species are variously called ecological engineers, habitat-formers or key species. Habitats formed by ecological engineers include some of the bioturbators we have discussed, for example, areas dominated by pits, burrows, galleries and holes or by heavy bulldozing of the surface sediments, or even by the faeces of head-down bottom-up deposit feeders.

However, there is another very important habitat type that has multiple effects on sedimentary and water column processes; relatively immobile structures that protrude above the sediment–water interface (often called 3-dimensional habitat structure). These can be structures created by infauna or epifauna such as the sand grain tubes or calcium deposits built by polychaetes, or they can be the plant or animal itself. Generally, these are intended to give better access to resources. In the case of plants, growth away from the sediment surface will give them better access to light and higher water flows that break down diffusional boundary layers and limit solute exchanges across blades. For others, such as suspension-feeding polychaete tube worms and bivalves such as mussels and oysters, the move off the sediment surface means that less of the particles encountered are inorganic (sediment) and thus the food is of higher quality. But for both plants and animals the consequences of protruding into the water column are exposure to higher velocities and the potential to affect hydrodynamic and sedimentary processes driven by alteration of the benthic boundary layer. This alteration has important implications for the benthos independent of any physiologically dependent processes and has been an important area of study for benthic ecologists.

The concepts of how structures alter and interact with boundary layer flows have been well established in the engineering/fluid mechanics literature for decades because of their relevance to bridge design, drag on ships' hulls etc. However, it is only comparatively recently that they have been applied to benthic ecology. In a series of elegant papers in the 1980s, Pete Jumars and Arthur Nowell and PhD student Jim Ekman cast the engineering literature into an ecological framework through an investigation of how individual and groups of tubes affected flow. This work demonstrated that with individual tubes high-momentum fluid is transported down toward the sediment surface, increasing the bed shear stress and destabilising the sediment (Carey, 1983; Eckman et al., 1981).

The next element came from considering the spatial structure of tube worm patches, seagrass and bivalve beds (Gambi et al., 1990; Green et al., 1998). As the areal density increases and patch length increases from the leading edge, a phenomenon known as skimming flow is induced. In essence the benthic boundary is now raised to the top of the structural elements, with a transition zone occurring between the structural elements—a zone of low flow and turbulence. This transition results in the structural elements moving from being sediment destabilisers to being sediment stabilisers where the sediment surface is protected from high bed shear stress upon turbulence (Figure 2.4). The density of structure required to induce skimming flow is a function of the vertical height, frontal area exposed to the flow and flexibility (especially important in terms of macrophytes)—a good description of this can be found in Eckman (1983) and Nepf and Koch (1999).

This work stimulated many more field and laboratory studies as technology for visualising and characterising flows has developed. In turn this has led to a better understanding of the implications. The alteration of the hydrodynamic regime in patches of sufficient size fundamentally alters benthic–pelagic coupling, changing settling of detrital matter to the seafloor and nutrient and oxygen fluxes into the water column. Reduced flows between structural elements allow the sedimentation of finer-grained particles and enhanced organic matter deposition (Bos et al., 2007). These changes in the fluxes of fine sediments can result in enhanced nutrient cycling within the bed. The alteration to the flow regime also has impacts on the settlement/recruitment of benthic invertebrates (as discussed in Chapter 7).

A final habitat type formed by interactions between organisms and sediments could be considered as a subset of the 3-dimensional habitat structure, although in this case the vertical dimension is very small: encrusting organisms and rhodolith beds. Small encrusting organisms such as coralline algae and surpulid worms can create a highly stable seafloor, inhibiting sediment resuspension, in high-flow and coarse sediment areas. This stability fundamentally changes the biogeochemistry of these areas.

Independent flow

Interactive flow

Skimming flow

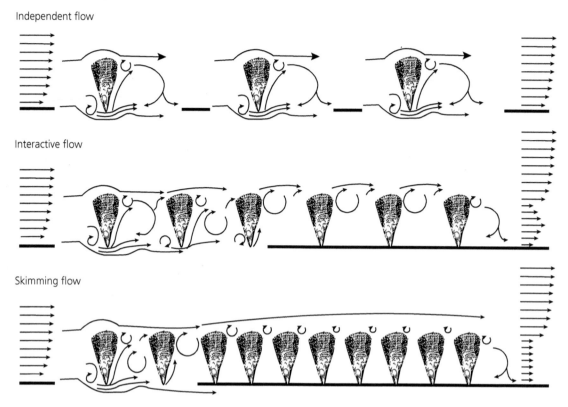

Figure 2.4 Effects of horse mussel density (*Atrina zealandica*) on benthic boundary layer flows. At low densities alterations to the flow caused by a mussel have dissipated before it encounters the next individual. At high densities skimming flow occurs where a new benthic boundary is established above the mussels and the zone between them becomes a low-flow area that promotes sedimentation. Figure modified from Friedrichs (2004).

2.7 Close out

The sedimentary environment that would develop under a world with no plants and animals is fundamentally different than that seen in most soft-sediment systems. Benthic species therefore strongly influence the balance between recycling vs sequestering of organic material. Increased oxic depth and enhanced particle transport and solute advection are common (Figure 2.5), but increased flux of organic material from the water column, changes in nutrient fluxes, increased carbon inputs and increased microphyte activity also occur. Simplistically, sediments with no fauna will not be as efficient at processing organic matter, meaning that more material is buried.

These differences mean that descriptions of sea-floor habitats based on environmental descriptors (such as sheltered intertidal mud flats and high current, coarse sediment) are of very limited value. These descriptions do not even provide a useful hierarchical categorisation of sedimentary processes such as organic matter remineralisation (see Figure 2.5), let alone ecological function or ecosystem services (see Chapter 9).

Many books describe biotic habitats based on different types of vegetation—e.g. seagrass, rhodolith beds, kelp, coralline algae, fringing vegetation—but we need a much wider consideration of biogenic habitats. However, differences in size, living position, feeding mode and mobility of the animals that inhabit various places result in a large potential range of non-vegetated habitats. Although we have suggested the possibility of additional categories based on degree of bioturbation, sediment stability (or lack of it) and erect

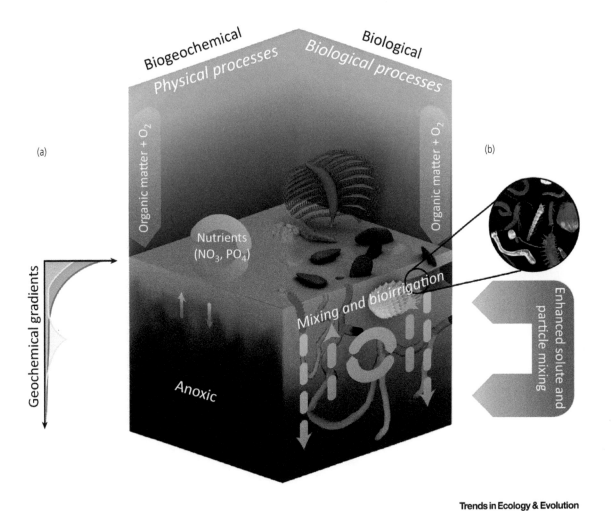

Figure 2.5 The world of the sediment is very different without animals. From Figure 1, Snelgrove et al. (2018). Reprinted from *Trends in Ecology & Evolution*, **33**, Snelgrove P V R, Soetaert K, Solan M, Thrush S F, Wei C-L, Danovaro R, Fulweiler R W, Kitazato H, Ingole B, Norkko A, Parkes R J, and Volkenborn H, Global Carbon Cycling on a Heterogeneous Seafloor, 96–105., Copyright (2017) with permissions from Elsevier Publication.

3-dimensional structure, different combinations of key species will result in different types of habitats that are important not just for ecological functioning but, at least as importantly, for biodiversity (see Chapters 6 and 7) and for sensitivity to and recovery from natural disturbances and human activities.

References

Aller R C (1982). The Effects of Macrobenthos on Chemical Properties of Marine Sediment and Overlying Water. In:

McCall P L and Tevesz M J S, eds. *Animal-Sediment Relations*, Springer, Boston, MA, pp. 53–102.

Aller R C (1988). Benthic Fauna and Biogeochemical Processes in Marine Sediment: The Role of Burrow Structures. In: Blackburn T H and Sorensen J, eds. *Nitrogen Cycling in Coastal Marine Environments*, J. Wiley and Sons, New York, pp. 301–38.

Aller R C (1994). Bioturbation and Remineralization of Sedimentary Organic Matter: Effects of Redox Oscillation. *Chemical Geology*, **114** (3–4), 331–45.

Bavestrello G, Arillo A, Calcinai B, Cattaneo-Vietti R, Cerrano C, Gaino E, Penna A and Sara M (2000). Parasitic Diatoms inside Antarctic Sponges. *The Biological Bulletin*, **198** (1), 29–33.

Bayne B L, Iglesias J I P, Hawkins A J S, Navarro E, Heral M and Deslouis-Paoli J M (1993). Feeding Behaviour of the Mussel, *Mytilus edulis*: Responses to Variations in Quantity and Organic Content of the Seston. *Journal of the Marine Biological Association UK*, **73**, 813–29.

Bonaglia S, Nascimento F A, Bartoli M, Klawonn I and Brüchert V (2014). Meiofauna Increases Bacterial Denitrification in Marine Sediments. *Nature Communications*, **5** (1), 1–9.

Bos A R, Bouma T J, de Kort G L and van Katwijk M M (2007). Ecosystem Engineering by Annual Intertidal Seagrass Beds: Sediment Accretion and Modification. *Estuarine, Coastal and Shelf Science*, **74** (1–2), 344–8.

Boudreau B P (1998). Mean Mixed Depth of Sediments: The Wherefore and the Why. *Limnology and Oceanography*, **43** (3), 524–6.

Cadee G C (1979). Sediment Reworking by the Polychaete *Heteromastus filiformis* on a Tidal Flat in the Dutch Wadden Sea. *Netherlands Journal of Sea Research*, **13**, 441–54.

Carey D A (1983). Particle Resuspension in the Benthic Boundary Layer Induced by Flow around Polychaete Tubes. *Canadian Journal of Fisheries and Aquatic Sciences*, **40** (S1), s301–8.

Coull B C (1999). Role of Meiofauna in Estuarine Soft-Bottom Habitats. *Australian Journal of Ecology*, **24** (4), 327–43.

De Boer W (2007). Seagrass–Sediment Interactions, Positive Feedbacks and Critical Thresholds for Occurrence: A Review. *Hydrobiologia*, **591** (1), 5–24.

Eckman J E (1983). Hydrodynamic Processes Affecting Benthic Recruitment. *Limnology and Oceanography*, **28**, 241–57.

Eckman J E, Nowell A R M and Jumars P A (1981). Sediment Destabilization by Animal Tubes. *Journal of Marine Research*, **39**, 361–74.

Fauchauld K and Jumars P A (1979). The Diet of Worms: A Study of Polychaete Feeding Guilds. *Oceanography and Marine Biology Annual Reviews*, **17**, 193–284.

Friedrichs M (2004). Flow-Induced Effects of Macrozoobenthic Structures on the Near-Bed Sediment Transport. *PhD thesis, University of Rostock, Rostock*.

Gambi M C, Nowell A R and Jumars P A (1990). Flume Observations on Flow Dynamics in *Zostera marina* (Eelgrass) Beds. *Marine Ecology Progress Series*, **61**, 159–69.

Grant J (1983). The Relative Magnitude of Biological and Physical Sediment Reworking in an Intertidal Community. *Journal of Marine Research*, **41**, 673–89.

Gray J S (1981). *The Ecology of Marine Sediments*, Cambridge University Press, Cambridge.

Green M O, Hewitt J E and Thrush S F (1998). Seabed Drag Coefficients over Natural Beds of Horse Mussels (*Atrina zelandica*). *Journal of Marine Research*, **56**, 613–37.

Hendriks I E, Bouma T J, Morris E P and Duarte C M (2010). Effects of Seagrasses and Algae of the Caulerpa Family on Hydrodynamics and Particle-Trapping Rates. *Marine Biology*, **157** (3), 473–81.

Herman P M J, Middelburg J J, VandeKoppel J and Heip C H R (1999). Ecology of Estuarine Macrobenthos. *Advances in Ecological Research*, **29**, 195–231.

Jones C G, Lawton J H and Shachak M (1994). Organisms as Ecosystem Engineers. *Oikos*, **69**, 373–86.

Jørgensen C B (1996). Bivalve Filter Feeding Revisited. *Marine Ecology Progress Series*, **147**, 287–302.

Kiorboe T and Mohlenberg F (1981). Particle Selection in Suspension-Feeding Bivalves. *Marine Ecology Progress Series*, **5**, 291–6.

Kregting L T, Hepburn C D, Hurd C L and Pilditch C A (2008). Seasonal Patterns of Growth and Nutrient Status of the Macroalga *Adamsiella chauvinii* (Rhodophyta) in Soft Sediment Environments. *Journal of Experimental Marine Biology and Ecology*, **360** (2), 94–102.

Kristensen E (2000). Organic Matter Diagenesis at the Oxic/Anoxic Interface in Coastal Marine Sediments, with Emphasis on the Role of Burrowing Animals. In: Liebezeit G, Dittmann S and Kröncke, I, eds. *Life at Interfaces and under Extreme Conditions*, Springer Netherlands, Dordrecht, pp. 1–24.

Kristensen E (2001). Impact of Polychaetes (Nereis spp. and *Arenicola marina*) on Carbon Biogeochemistry in Coastal Marine Sediments. *Geochemical Transactions*, **2** (12), 92–103.

Kristensen E, Penha-Lopes G, Delefosse M, Valdemarsen T, Quintana C O and Banta G T (2012). What Is Bioturbation? The Need for a Precise Definition for Fauna in Aquatic Sciences. *Marine Ecology Progress Series*, **446**, 285–302.

Krueger F (1964). Experiments Concerning the Dependence of Respiration of *Arenicola marina* (Annelides Polychaeta) on Size and Temperature. *Helgoland Wissenschaftliche Meeresunterschungen*, **10**, 38–63.

Levin L, Blair N, DeMaster D, Plaia G, Fornes W, Martin C and Thomas C (1997). Rapid Subduction of Organic Matter by Maldanid Polychaetes on the North Carolina Slope. *Journal of Marine Research*, **55**, 1–17.

Lohrer A M, Thrush S F and Gibbs M M (2004). Bioturbators Enhance Ecosystem Performance via Complex Biogeochemical Interactions. *Nature*, **431**, 1092–5.

Lopez G, Taghon G and Levinton J (1989). *Ecology of Marine Deposit Feeders*, Springer-Verlag, New York.

Middelburg J J (2017). To the Bottom of Carbon Processing at the Seafloor: Towards Integration of Geological, Geochemical and Ecological Concepts (Vladimir Ivanovich Vernadsky Medal Lecture). *EGU General Assembly Conference Abstracts*, 2283.

Needham H R, Pilditch C A, Lohrer A M and Thrush S F (2012). Density and Habitat Dependent Effects of Crab

Burrows on Sediment Erodibility. *Journal of Sea Research*, **76**, 94–104.

Nepf H M and Koch E W K (1999). Vertical Secondary Flows in Submersed Plant-Like Arrays. *Limnology and Oceanography*, **44** (4), 1072–80.

Norkko A, Hewitt J E, Thrush S F and Funnell G A (2001). Benthic–Pelagic Coupling and Suspension Feeding Bivalves: Linking Site-Specific Sediment Flux and Biodeposition to Benthic Community Structure. *Limnology & Oceanography*, **46**, 2067–72.

Reise K (1983). Biotic Enrichment of Intertidal Sediments by Experimental Aggregates of the Deposit-Feeding Bivalve *Macoma balthica*. *Marine Ecology Progress Series*, **12** (3), 229–36.

Rhoads D C (1963). Rates of Sediment Reworking by *Yoldia limatula* in Buzzards Bay, Massachusetts and Long Island Sound. *Journal of Sedimentary Petrology*, **33**, 723–7.

Rhoads D C (1974). Organism-Sediment Relations on the Muddy Seafloor. *Oceanography and Marine Biology: An Annual Review*, **12**, 263–300.

Rhoads D C and Boyer L F (1982). The Effects of Marine Benthos on Physical Properties of Sediments. A Successional Perspective. In: McCall P L and Tevesz M J S, eds. *Animal-Sediment Relations*, Springer, Boston, MA, pp. 3–52.

Rhoads D C and Young D K (1970). The Influence of Deposit-Feeding Organisms on Sediment Stability and Community Trophic Structure. *Journal of Marine Research*, **28**, 150–78.

Rosa M, Ward J E and Shumway S E (2018). Selective Capture and Ingestion of Particles by Suspension-Feeding Bivalve Molluscs: A Review. *Journal of Shellfish Research*, **37** (4), 727–46.

Safi K A, Hewitt J E and Talman S (2007). The Effect of High Inorganic Seston Loads on Prey Selection by the Suspension-Feeding Bivalve, *Atrina zelandica*. *Journal of Experimental Marine Biology and Ecology*, **344**, 136–48.

Timmermann K, Banta G T and Glud R N (2006). Linking *Arenicola marina* Irrigation Behavior to Oxygen Transport and Dynamics in Sandy Sediments. *Journal of Marine Research*, **64** (6), 915–38.

Volkenborn N, Polerecky L, Wethey D S and Woodin S A (2010). Oscillatory Porewater Bioadvection in Marine Sediments Induced by Hydraulic Activities of *Arenicola marina*. *Limnology and Oceanography*, **55** (3), 1231–47.

Volkenborn N, Robertson D M and Reise K (2009). Sediment Destabilizing and Stabilizing Bio-Engineers on Tidal Flats: Cascading Effects of Experimental Exclusion. *Helgolander Meeresuntersuchungen*, **63**, 27–35.

Wethey D S and Woodin S A (2005). Infaunal Hydraulics Generate Porewater Pressure Signals. *The Biological Bulletin*, **209** (2), 139–45.

Woodin S A, Volkenborn N, Pilditch C A, Lohrer A M, Wethey D S, Hewitt J E and Thrush S F (2016). Same Pattern, Different Mechanism: Locking onto the Role of Key Species in Seafloor Ecosystem Process. *Scientific Reports*, **6**, 26678.

Wooster M K, McMurray S E, Pawlik J R, Morán X A and Berumen M L (2019). Feeding and Respiration by Giant Barrel Sponges across a Gradient of Food Abundance in the Red Sea. *Limnology and Oceanography*, **64** (4), 1790–801.

Yahel G, Sharp J H, Marie D, Häse C and Genin A (2003). In Situ Feeding and Element Removal in the Symbiont-Bearing Sponge *Theonella swinhoei*: Bulk DOC Is the Major Source for Carbon. *Limnology and Oceanography*, **48** (1), 141–9.

CHAPTER 3

Disturbance, patches and mosaics

3.1 Introduction

Disturbance plays an important role affecting the structure and function of benthic communities. Individual disturbance events can decimate local communities, but also free up resources, provide space for colonisation and change ecosystem function. How individual disturbance events affect the seafloor will depend on both the scale of the disturbance event and whether we are interested in the ecological consequences at the scale of the disturbed patch or across the broader landscape/ecosystem.

In marine sediments, early research on disturbance by Ralph Johnson introduced the mosaic concept of patches of the seafloor in different stages of recovery from disturbance (Johnson, 1970). The interest in disturbance ecology grew because research on small disturbance events, often associated with the actions of predators feeding in sediments, was easy to replicate experimentally (Dayton et al., 1974; Levin, 1984; Van Blaricom, 1982). These experimental studies were complemented by the analysis of the broader-scale patterning of the seafloor that led to the analysis of spatial and temporal gradients of disturbance (Pearson & Rosenberg, 1978; Rhoads, 1974). These studies changed our understanding of seafloor communities. No longer was our view of the seafloor one of stability and consistency; now, disturbance was creating patches and taking us to a more dynamic view of seafloor communities and habitats.

Advances in our understanding of the ecological patterns and processes associated with disturbance events have always been strongly linked to human impacts. Concerns about the impacts of trawling, dredging, removal of shellfish beds, mining and oil and gas extraction, sedimentation events, algal blooms and hypoxia/anoxia have all been informed by the application of disturbance ecology. Today, insights from disturbance ecology are pervasive in impact assessment, risk assessment, cumulative effects studies, marine management and the development of scenarios of future change.

Disturbance ecology has never just been about the biological, physical or chemical factors that disturb the seafloor; the description of the event has always been linked to the subsequent recovery processes. In the 1970s–90s much of the research focussed on defining patch dynamics and recovery processes; now, questions have moved on to considering changes in ecosystem function and the potential for cumulative effects to radically change seafloor ecologies. Links between impact and recovery and the potential for feedbacks between the two have brought us to begin to understand how these relationships can shape the structure and function of vast tracts of the seafloor and lead to regime shifts. In current times, when concerns of biodiversity loss and its consequences are growing, understanding disturbance effects is not only high on the agenda, but a key to insightful management of soft-sediment habitats.

3.2 Disturbance events

Individual disturbance events can range from small-scale phenomena like the bulldozing of the sediment by urchins and bivalves or the probing, biting and digging of predators, to the gouging of the seafloor by icebergs or the decomposition of

Ecology of Coastal Marine Sediments: Form, Function, and Change in the Anthropocene. Simon F. Thrush, Judi E. Hewitt, Conrad A. Pilditch and Alf Norkko,
Oxford University Press (2021). © Simon Thrush, Judi Hewitt, Conrad Pilditch, and Alf Norkko.
DOI: 10.1093/oso/9780198804765.003.0003

Figure 3.1 Different types of seafloor disturbances. (A) Sedimentation event; (B) ray pit; (C) localized sediment event; (D) ripples generated by waves; (E) Shellfish washed up on a beach by a storm; (F) bulldozed track from a bivalve; (G) solitary dead *Atrina* in an anoxic zone; (H) starfish digging a pit.

seaweed wrack, to large-scale impacts associated with storms, earthquakes and volcanic eruptions and changes in ocean circulation. Disturbances are discrete events, although to some extent this is just a matter of time scale. We can consider a ray digging a hole in a sandflat over a few minutes and a slow change in climate or pollution that may build over decades both to be disturbance events (Figure 3.1).

Disturbances not only directly impact the organisms that live on the seafloor but also physically or chemically alter their habitats. The pit left behind by a feeding ray will not only contain fewer prey items, it will also modify the hydrodynamics, possibly accumulating fine organic-rich sediments. This grand act of bioturbation will have changed sediment and pore water characteristics. Mass ray feeding can turn over a large proportion of the top 20 cm of sediment. In general, the removal of animals and changes in the environment provide opportunities for other species. These may be species unaffected by disturbance that then increase in abundance or new species that arrive at the disturbed site. Individual species may be differentially susceptible to disturbance because of their tolerance to changing conditions. This relates to the biological traits of species associated with physiological tolerances (e.g. hypoxia), sensitivity to physical damage (e.g. trawling) or lifestyle (ability to burrow out of trouble or move away temporarily to a refuge habitat). Susceptibilities of individual species may also be life-stage- or size-dependent.

Through the removal of individuals or whole communities from patches of the seafloor and changes to the physical nature of sediments, disturbance is often considered to be a density-independent source of mortality. However, abundance may affect either the risk of disturbance or the susceptibility to disturbance. For example, rays feeding on sandflats, like many predators, target high-density patches (Hines et al., 1997), and shellfish at high abundance may be less susceptible to elevated suspended sediment concentrations than sparsely distributed populations (Coco et al., 2006).

Changes in biogenic structures, sediment grain size and biogeochemistry can have significant legacy effects influencing recovery processes and long-term community dynamics and functions. Although

individual events might be small, over large areas repeated disturbance events can have big impacts. In fact, when jaws evolved in the Precambrian many animals started to burrow into the seafloor to escape predation; this introduced the processes of bioturbation and bioirrigation that have mobilised nutrients and enhanced productivity and diversity over the Phanerozoic (Martin, 1996).

3.3 Disturbance regimes

The intensity of disturbance combined with information on the size of the area directly impacted and the frequency of disturbance events define a disturbance regime. This is a useful way of connecting the effects of individual disturbance events to provide insight into what is happening over large space and time scales. Disturbance regimes are particularly useful in providing insights into the ways humans alter seafloor ecosystems because often what our actions do is change the disturbance regime. For example: in regions unaffected by storms, large-scale physical disturbance of the seafloor was infrequent before the advent of trawl and dredge fisheries; nutrient run-off has increased the extent and frequency of hypoxic events; and climate change, at least in some parts of the world, has increased the frequency of storms. Humans have shifted natural disturbance regimes as we discuss later in Chapter 11.

There is often a power–function relationship linking frequency and extent of disturbance. Essentially, small disturbances occur exponentially more frequently than large ones (see Bak, 1996). Consequently, many small disturbance events will have different effects on seafloor ecosystems than occasional large ones. Thinking about how and why different disturbance regimes affect the structure and function of the seafloor can be better informed by focussing on these relationships rather than simple categorisation of disturbance events into big and small or quick and slow. Such classifications tend to cloud questions as much as inform them. For example, Bender et al. (1984) classified disturbances into fast (press disturbance) and slow (pulse disturbance). This classification was thought to help differentiate short events that allow the system to return to equilibrium and longer disturbances that might drive the system into an alternate state (Ives & Carpenter, 2007). But small and frequent disturbances can be as important in defining state as large and long-lasting disturbances. Also, a specific disturbance may affect microbial, meiofaunal, macrofaunal and megafaunal communities in very different ways. The time scales of response of these different organisms can be very diverse and consequently disturbance events may be very discrete for one group of organisms but represent a long-term change for another. This has led to some argument about 'what is a disturbance' for different sediment-dwelling organisms (Allison & Martiny, 2008) when individual species or size groupings of organisms respond to the same disturbance in different ways. Thinking about how responses to disturbance vary across disturbance regimes is perhaps more profitable than simple classifications.

Another approach to search for generality in the role of disturbance effects is looking at the relationship between disturbance and diversity. This generality has often focussed on the frequency of disturbance events. For soft sediments, the most commonly referred to (but rarely tested) is the intermediate disturbance hypothesis. This hypothesis grew out of research on terrestrial plants (Grime, 1973) and rocky shores and coral reefs (Connell, 1977)—systems that exhibit strong competition for space. The idea is that as disturbance creates holes in the area occupied by the competitive dominant, new space can be colonised by other species. So not enough disturbance and the diversity is low and dominated by the space dominant; too much disturbance and the space dominant is lost and almost everything is disturbed too frequently for populations to flourish. Somewhere in the middle—at intermediate disturbance—the hypothesis says species richness should be highest. This hypothesis has been used to argue that human disturbance of the seafloor is a good thing, enhancing diversity. Unfortunately for these proponents, this hypothesis has not found support everywhere. The scale of disturbances created by human activities is often markedly different from that of natural disturbances and ecosystem function is not simply related to species richness (see Chapters 9 and 11).

Application of general theories outside their original assumptions, especially when not empirically tested, is highly problematic. Strong competition for space occupancy is not the norm on the seafloor, in the same way it can be in terrestrial plant and rocky shore systems, because sediments are a much more three-dimensional system and most of the animals have some mobility. The way we sample and thus view soft sediments also emphasises the importance of multiple trophic levels and here too the theory fails (Wootton, 1998). In fact, this parabolic relationship between disturbance and diversity is only one of a whole suite of possible relationships (Miller et al., 2011) and has been subject to extensive criticism. We still have much to learn about the mechanisms that lie behind the relationship between disturbance and diversity (Fox, 2013). Theory is important in advancing the science, but we need to be careful not to misapply concepts or use theories without strong empirical tests.

Another example of generality developed out of 1970s research on both sides of the Atlantic focussed on how benthic communities responded to disturbance (Figure 3.2). In Europe, Tom Pearson and Rutger Rosenberg were developing their ideas by looking at spatial gradients running away from areas of high organic matter deposition, while in Long Island Sound (USA) Don Rhoads and colleagues were looking at the recovery of soft-sediment organisms from anoxia. The resultant conceptual models were very similar and have been exceedingly influential in how we think about disturbance to the seafloor. Concepts such as the increase in opportunistic species and gradients in diversity and abundance have been central to many of the 'health' indicators that have been developed and applied in many environmental management contexts. However, both these studies considered disturbance associated with organic enrichment and anoxia and the response of communities living in muddy sediments. Pearson and Rosenberg (1978) fully acknowledge that their model might not apply to more hydrodynamically active and sandy sediment conditions and indeed its application to different types of habitats and systems has yielded variable results (Thrush & Whitlatch, 2001).

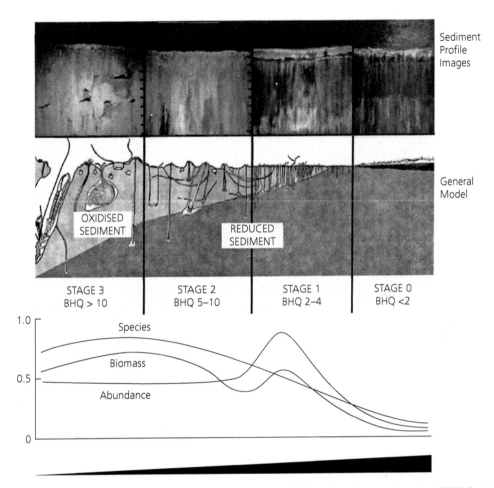

Sediment
Profile
Images

General
Model

OXIDISED
SEDIMENT

REDUCED
SEDIMENT

STAGE 3
BHQ > 10

STAGE 2
BHQ 5–10

STAGE 1
BHQ 2–4

STAGE 0
BHQ <2

Species

Biomass

Abundance

Figure 3.2 Conceptual response of marine benthic habitats and fauna to oxygen deficiency. From Nilsson and Rosenberg (1994). Permission granted by Marine Ecology Progress Series, © Inter-Research 2000. BHQ is an index used to measure benthic health along the enrichment gradient.

3.4 Recovery processes

Once the disturbance has abated, organisms can start to recolonise. The time to recover will be influenced by the type of disturbance and its location, but even small experimental plots (1 m to tens of metres) on dynamic sandflat habitats can take over a year to recover (Beukema et al., 1999; Norkko et al., 2006; Thrush et al., 1996, 2008; Volkenborn & Reise, 2006). The recovery process can be quite complicated and affected both by factors that supply colonists to the disturbed patch

and by factors that occur within the disturbed patch (Figure 3.3).

Within the patch many of the interactions that occur between organisms and their physico-chemical environment are interconnected. A series of papers produced by PhD students, post-docs and their supervisors working together on different aspects of recovery illustrate this point (Montserrat et al., 2008; Rossi et al., 2004; Van Colen et al., 2008, 2010). Here we can see (Figure 3.4) a burst in microphytobenthos causing suspended muddy sediments to stick in the plot. Mud snails crawled

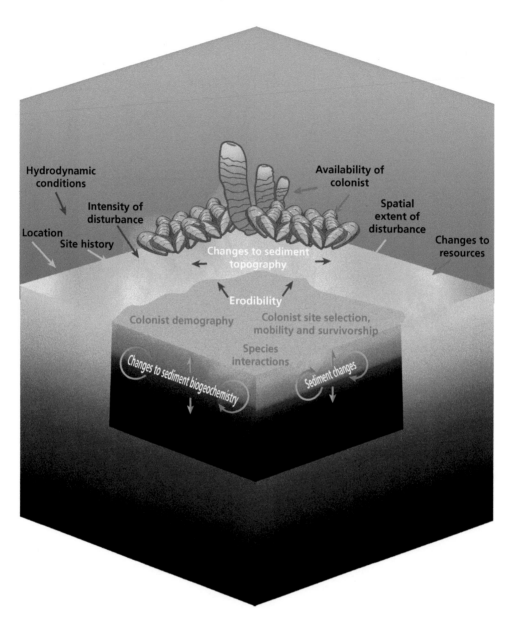

Figure 3.3 Multiple factors influencing recovery in and out of a disturbed patch, based on Thrush and Whitlatch (2001) with mussel and sponge illustrations by Dieter Tracey and Caroline Wicks, IAN Image Library (https://ian.umces.edu/imagelibrary/).

into the plots within a few days to graze on these microphytes. Next came the tube-building worms and the sub-surface deposit feeders, the latter disturbing the sediment surface. The grazing and sediment disturbance allowed the accumulated fine sediment to erode from the plots. The study shows how the time scales of different events occurring within the plots are interrelated with macrofaunal

recovery. The initial bloom of microphytobenthos following disturbance occurs because of rapid colonisation of the microphytes associated with sediment bedload transport, rapid growth and the absence of grazing pressure in the recently disturbed plots. These types of interrelationships have been demonstrated in other disturbance experiments. However, there are system-specific

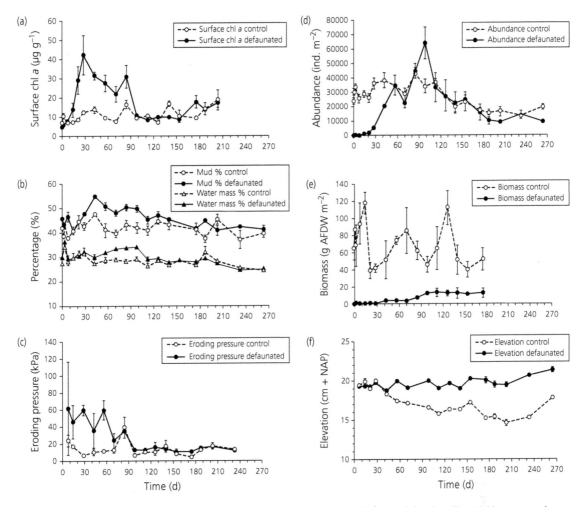

Figure 3.4 Changes over time in biotic and abiotic parameters in ambient conditions and defaunated plots. From Figure 2, Montserrat et al. (2008). Reprinted under Creative Commons Attribution 4.0 International (CC BY 4.0) licence. AFW: ash-free dry weight; NAP: zero reference for elevation in the Netherlands (Normal Amsterdam Level).

responses—in non-eutrophic systems where the microphytes need the macrofauna to release nutrients from the sediment, there is no microphyte bloom but a steady increase associated with colonisation by bioturbators and biodiffusers.

Recovery processes are often referred to as succession, disturbance-recovery dynamics or community assembly. One phenomenon that can occur in the early stages of recovery is an explosion in the abundance of opportunistic species. These are classified as species that early in the recovery process achieve higher density in the disturbed plots than in the adjacent undisturbed sediment. These oppor-

tunists are thought to ameliorate unfavourable conditions, making the disturbed plot suitable for later colonists. Although common in disturbances involving organic enrichment of the sediment, opportunistic responses are not always observed. In a multi-site experiment that looked at recovery of anoxic plots, opportunistic recruitment events were not drivers of recovery. A massive recruitment of some species occurred but only 203 days after the start of the recovery and the abundance of these individuals largely decreased by the end of the experiment, regardless of location in the estuary or hydrodynamics (De Juan et al., 2014).

The challenges to understanding recovery involve integrating contributing factors: different modes of colonisation of the disturbed area, which can be affected by the timing of disturbance events; the scale of disturbance and the location of a disturbed patch in the seafloor's mosaic of patches; the potential for species interactions within the patch which may facilitate or inhibit other colonists; and legacy effects associated with habitat change. Despite all the possible permutations, some common patterns emerge when the scale of recruitment is mismatched with the recovery potential of the disturbed patch. These patterns include the loss of large and long-lived species, the fragmentation of habitats dominated by 'late recovery stage' communities, and eventually the homogenisation of seafloor habitats as the system becomes dominated by small, mobile and rapid-growing species.

The spatial scale of disturbance can influence recovery and this can be especially important when using small-scale experimental studies to predict and assess human impacts. Much of our understanding of the basic ecology of disturbance-recovery processes in marine sediments comes from manipulative field experiments. These are a good way to understand the role of different processes, but it may not be a good idea to directly scale-up to predict effects and recovery rates in large-scale human disturbances (see Chapter 4). Experiments that have been conducted to investigate recovery processes in different-size plots often reveal insightful but counter-intuitive results (Norkko et al., 2006). For example, 9 months after the start of an experiment that defaunated plots of 0.2, 0.8 and 3.2 m² there were still significant differences in the community composition (Thrush et al., 1996)—and these were relatively small differences in disturbance size. If we look at the MDS ordination plot of how the community varied over the course of the experiment (Figure 3.5) we can see the lack of overlap at the end of the experiment and the much wider variability in community composition during the recovery process that occurred with increasing plot size.

Whereas most disturbance recovery experiments focus on colonisation, the experiment described

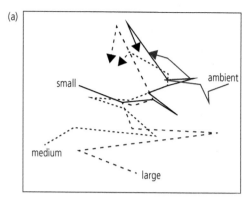

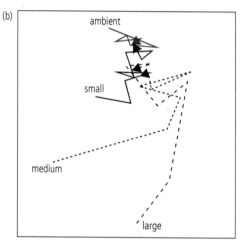

Figure 3.5 Recovery trajectories (a) species abundances and (b) species presence/absence over time were dependent on the size of the disturbed plots. Purple = ambient sediment; black solid = smallest plot; black dashed = medium sized; black dotted = largest plot. Adapted from Thrush et al. (1996).

above taught us a lesson on the importance of emigration from the disturbed plots. The defaunation had killed off a dense tube mat of spionid polychaetes that stabilised the sediment from the action of waves. With the tube mat gone it was hard for the colonising animals to not get eroded out of the plots and as the scale of the disturbance increased this problem increased. This experiment demonstrated the importance of the net effect of immigration and emigration and the role of tube mats in mediating sediment stabilisation.

The role of animals in mediating habitat change often becomes clearer as we increase the scale of disturbance. The importance of biotic habitat modification and indirect effects increasing with spatial or temporal scale is a message that comes through the literature. Using 400-m² exclusion plots to exclude *Arenicola marina* from sandflats in the Wadden Sea Volkenborn and Reise (2006) showed how surface sediment destabilisation by adult *Arenicola marina* had negative effects on the nereid *Nereis diversicolor* and juvenile conspecifics—but a facilitatory effect on the subsurface-dwelling orbinid *Scoloplos* cf. *armiger*. The strength of these relationships, especially between *Arenicola* and *Nereis*, varied between years and between plots at slightly different tidal elevations—again highlighting how effects can change with scale. This experiment is also interesting because it highlights that not everything has to die as a result of a disturbance; some species can be left behind. In this case subsurface nets were used to restrict the density of the large deep-dwelling *Arenicola*. The question of who is left behind if not all organisms are eliminated is interesting, since potential 'priority effects', i.e. whether the identity of the first colonists matters and to what extent they may dictate the successional process, are poorly known and this affects our ability to generalise successional models.

Contrasting the results of a field experiment and a gradient design, Norkko et al. (2006) highlighted how the role of opportunistic species in recovery became clearer as the scale of disturbance increased because of influences on the level of resources available to colonists and the degree to which species were released from competitive interactions. These trends are interesting as we might expect the role of biology to increase as the recovery process proceeds and larger organisms and a higher diversity of species (and thus functional groups) are found. We might expect physical and chemical processes to dominate in highly disturbed conditions where few animals survive, but analysis of how animals contribute to ecosystem processes shows that they can be especially important in these conditions.

3.5 Dispersal

The dispersal of organisms is critical to supplying colonists to disturbed patches. Evidence is growing

that we should not think of soft sediments as completely open systems where any species can make it to colonise a disturbed site. Individual species each have their own dispersal characteristics and one of the limitations of full recovery is that many of the mature successional stage species have very restricted dispersal and are consequently slow to colonise (Grantham et al., 2003; Kinlan & Gaines, 2003). Many soft-sediment species have different modes of dispersal that are linked to the life stage of the organism and these age-dependent dispersal processes can have very different length scales. Larval dispersal will depend on water currents and may move organisms over kilometres of the seafloor. Once the organism has initially settled, it may undergo a second phase of post-larval dispersal. This again may have the animal moving long distances with water currents or have it moving over tens to hundreds of metres of sediment with bedload transport. Finally, adult life stages may be relatively sedentary or actively crawl over the sediment surface or swim in the water column. As the areal extent of disturbance increases, recolonisation depends on the ability of colonists to travel longer distances which restricts the potential pool of colonists.

Different behaviours and the relative importance of active verse passive transport processes can vary with habitat, life stage and species. This can affect what life stage of an organism is likely to colonise a large or a small disturbed patch. Larval and juvenile stages may settle en masse whereas older animals are likely to colonise at lower abundances. Nevertheless, because they arrive as adults, they may be better able to tolerate unfavourable conditions and out-compete competitors. Different sources of colonist supply and proximity of source populations link to the scale of disturbance in terms of both extent and frequency. Defining the length scale of dispersal and tracking the movement of individuals are difficult and we often resort to hydrodynamic particle dispersal models to identify a cloud of potential colonisers. This might be tractable for individual species but it becomes increasingly difficult as we start to consider the recolonisation of communities that contain many species (Jacobson & Peres-Neto, 2010).

3.6 From patch dynamics to meta-communities

The combination of factors that occur within a disturbed patch, the conditions on the seafloor in the vicinity of the patch and the dispersal processes that connect the two can be characterised as meta-community dynamics (Leibold et al., 2004). Meta-communities are characterised by the importance of both dispersal and in-patch processes—they are different from open systems where all species can disperse over the entire landscape and closed systems where dispersal out of the patch is not possible. There are many theoretical studies of meta-communities that have provided some interesting insights into the dynamics of communities across landscapes, but empirical studies, especially on the seafloor, are rare. Regardless, meta-communities help us think about how a series of small disturbances can change the seafloor landscape as a result of cumulative effects.

The critical point is that disturbance can affect diversity, as a consequence of both the loss of organisms in a disturbed patch and the changes in habitat features that can be important in enhancing species richness and functional diversity across the seafloor. We can think of this as a feedback process linking recovery of individual disturbed patches to what is happening at the scale of the seafloor landscape. This links backs to Johnson's 'mosaic of patches' view of the seafloor. As more of the seafloor is represented by disturbed patches there will be less of the seafloor containing patches of mature communities that can supply colonists to progress the recovery of the disturbed areas. These spatial dynamics are affected by the nature of the seafloor community and the time scales of recovery. When the seafloor is dominated by fast-growing opportunistic species patch recovery rates will be fast, but if the seafloor is dominated by large and slow-growing species then recovery will be slow. These space and time scales interact. If the time scales for complete recovery of a patch are slow and the frequency of disturbance means that more patches are opened up by disturbance than are fully recovered, then the seafloor will lose mature community members, especially when these are large, long-lived and structure-forming species. In other words,

there is a feedback between disturbance and the ecological response to disturbance at the landscape scale (Hughes et al., 2007).

Recognising feedback processes can give us valuable insight into the resilience of the seafloor to changes in the disturbance regime. The seafloor is heterogeneous and we might expect the strength of these feedbacks to vary across habitat types. Given the difficulties in defining dispersal accurately for even one species, we have a problem when trying to assess whole communities and we need to think about surrogate measures for community connectivity. There are multiple factors that might be useful clues about potential connectivity. These include the spatial pattern of habitat types and levels of diversity on the seafloor. Also there are indicators based on the hydrodynamics that transport potential colonists and beta-diversity is a community measure that links local richness to the regional species pool (see Chapter 6). If we can define these surrogates for connectivity then experimental disturbances across the landscape can provide a test of these concepts and provide empirical evidence that cumulative effects are a real phenomenon on the seafloor. Conducting the experiment at multiple sites allows us to assess if there are differences in recovery across the landscape. If there are differences in recovery rates, then we can develop statistical models to assess the importance of connectivity surrogates in explaining the differences (Thrush et al., 2008). These types of experiments can further develop these ideas by investigating the importance of the timing of disturbance (e.g. in periods of high or low larval recruitment) and manipulation of habitat complexity (Thrush et al., 2013). Experiments such as these are not simple to do, but they are important because they emphasise that disturbed patches do not recover in isolation and individual disturbance events cannot simply be added up to define cumulative effects.

Whereas conceptual models of disturbance and recovery have informed our thinking about seafloor ecological dynamics for many decades, there are fewer modelling studies that try to link processes involved in the recovery process and across the seafloor. This is not surprising given the context-dependent empirical results and the large number of factors that can influence recovery and patch

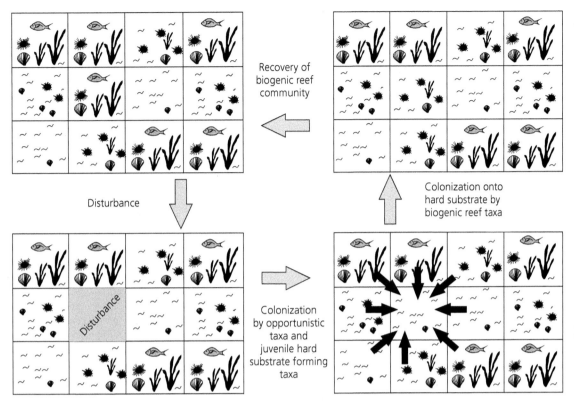

Figure 3.6 Conceptual model of a mosaic of habitat patches formed by disturbance and recovery. Figure 1 from Lundquist et al. (2006). Reprinted under Creative Commons Attribution 4.0 International (CC BY 4.0) licence.

dynamics. However, we might expect more attention in this area, given the importance of disturbance ecology in underpinning applied research on many types of human impacts.

Simple models do have value in helping us understand the interactions and relative importance of processes under different conditions. One example of this which builds on the empirical research on meta-community dynamics is provided by Lundquist et al. (2006). This model is illustrated in Figure 3.6. Basically, the seafloor is considered as a mosaic of tiles (model cells) with different tiles at different recovery stages following disturbance. The recovery stages are very simplified and range from dominance by small fast-growing species to dominance by large long-lived organisms. Dispersal rules for different successional stages set how many tiles a dispersal stage can cross. Even these very simplified situations highlight how interactions between changes in the disturbance regime, habitat

connectivity and the configuration of the landscape shape the seafloor habitat structure in quite complex ways.

3.7 Close out

We have learned a lot about soft-sediment ecology from studying disturbance. This has provided insight into how the seafloor is spatially structured, how habitat heterogeneity and diversity can vary and how the interactions between disturbance and recovery can affect temporal dynamics—including the potential for regime shifts. This fundamental knowledge has been applied to the development of health indicators, ecological impact assessment and risk analysis, marine conservation and ecologically sustainable resource use. There are still lots of habitats and ecosystems from which we do not have good information on recovery dynamics or the potential for interactions between different disturbance agents. We also need to better link changes in community structure and sediment characteristics to changes in ecosystem function so that we can understand the consequences of disturbance at both patch and seafloor scales. Nevertheless, the better we are at defining recovery, through empirical measurement and an understanding of the under- lying processes, the better we will be at moving beyond expert judgement to parameterise models that predict community recovery for ecosystem- based management (Teck et al., 2010).

References

Allison S D and Martiny J B H (2008). Resistance, Resilience, and Redundancy in Microbial Communities. *Proceedings of the National Academy of Sciences of the United States of America*, **105**, 11512–19.

Bak P (1996). *How Nature Works: The Science of Self-Organised Criticality*, Copernicus, New York.

Bender E A, Case T J and Gilpin M E (1984). Perturbation Experiments in Community Ecology: Theory and Practice. *Ecology*, **65**, 1–13.

Beukema J J, Flach E C, Dekker R and Starink M (1999). A Long-Term Study of the Recovery of the Macrozoobenthos on Large Defaunated Plots on a Tidal Flat in the Wadden Sea. *Journal of Sea Research*, **42** (3), 235–54.

Coco G, Thrush S F, Green M O and Hewitt J E (2006). Feedbacks between Bivalve Density, Flow, and Suspended Sediment Concentration on Patch Stable States. *Ecology*, **87**, 2862–70.

Connell J H (1977). Diversity in Tropical Rainforests and Coral Reefs. *Science*, **199**, 1302–10.

Dayton P K, Robilliard G A, Paine R T and Dayton L B (1974). Biological Accommodation in the Biological Community at Mcmurdo Sound. *Ecological Monographs*, **44**, 105–28.

De Juan S, Thrush S F, Hewitt J E, Halliday J and Lohrer A M (2014). Cumulative Degradation in Estuaries: Contribution of Individual Species to Community Recovery. *Marine Ecology Progress Series*, **510**, 25–38.

Fox J W (2013). The Intermediate Disturbance Hypothesis Should Be Abandoned. *Trends in Ecology & Evolution*, **28**, 86–92.

Grantham B A, Eckert G L and Shanks A L (2003). Dispersal Potential of Marine Invertebrates in Diverse Habitats. *Ecological Applications*, **13** (1), S108–16.

Grime J P (1973). Competitive Exclusion in Herbaceous Vegetation. *Nature*, **242**, 344–7.

Hines A H, Whitlatch R B, Thrush S F, Hewitt J E, Cummings V J, Dayton P K and Legendre P (1997).

Nonlinear Foraging Response of a Large Marine Predator to Benthic Prey: Eagle Ray Pits and Bivalves in a New Zealand Sandflat. *Journal of Experimental Marine Biology and Ecology*, **216**, 211–28.

Hughes A R, Byrnes J E, Kimbro D L and Stachowicz J J (2007). Reciprocal Relationships and Potential Feedbacks between Biodiversity and Disturbance. *Ecology Letters*, **10**, 849–64.

Ives A R and Carpenter S R (2007). Stability and Diversity of Ecosystems. *Science*, **317**, 58–62.

Jacobson B and Peres-Neto P R (2010). Quantifying and Disentangling Dispersal in Metacommunities: How Close Have We Come? How Far Is There to Go? *Landscape Ecology*, **25**, 495–507.

Johnson R G (1970). Variation in Diversity within Benthic Marine Communities. *American Naturalist*, **104**, 285–300.

Kinlan B P and Gaines S D (2003). Propagule Dispersal in Marine and Terrestrial Environments: A Community Perspective. *Ecology*, **84**, 2007–20.

Leibold M A, Holyoak M, Mouquet N, Amarasekare P, Chase J M, Hoopes M F, Holt R D, Shurin J B, Law R, Tilman D, Loreau M and Gonzalez A (2004). The Metacommunity Concept: A Framework for Multi-Scale Community Ecology. *Ecology Letters*, **7**, 601–13.

Levin L A (1984). Life History and Dispersal Patterns in a Dense Infaunal Polychaete Assemblage: Community Structure and Response to Disturbance. *Ecology*, **65**, 1185–200.

Lundquist C J, Thrush S F, Hewitt J E, Halliday J, MacDonald I and Cummings V J (2006). Spatial Variability in Recolonisation Potential: Influence of Organism Behaviour and Hydrodynamics on the Distribution of Macrofaunal Colonists. *Marine Ecology Progress Series*, **324**, 67–81.

Martin R E (1996). Secular Increase in Nutrient Levels through the Phanerozoic: Implications for Productivity, Biomass and Diversity of the Marine Biosphere. *Palaios*, **11**, 209–19.

Miller A D, Roxburgh S H and Shea K (2011). How Frequency and Intensity Shape Diversity–Disturbance Relationships. *Proceedings of the National Academy of Sciences of the United States of America*, **108**, 5643–8.

Montserrat F, Van Colen C, Degraer S, Ysebaert T and Herman P M J (2008). Benthic Community-Mediated Sediment Dynamics. *Marine Ecology Progress Series*, **372**, 43–59.

Nilsson H C and Rosenberg R (1994). Hypoxic Response of Two Marine Benthic Communities. *Marine Ecology Progress Series*, **115** (3), 209–17.

Norkko A, Rosenberg R, Thrush S F and Whitlatch R B (2006). Scale- and Intensity-Dependent Disturbance Determines the Magnitude of Opportunistic Response. *Journal of Experimental Marine Biology and Ecology*, **330** (1), 195–207.

Pearson T H and Rosenberg R (1978). Macrobenthic Succession in Relation to Organic Enrichment and Pollution of the Marine Environment. *Oceanography and Marine Biology: An Annual Review*, **16**, 229–311.

Rhoads D C (1974). Organism-Sediment Relations on the Muddy Seafloor. *Oceanography and Marine Biology: An Annual Review*, **12**, 263–300.

Rossi F, Herman P M J and Middelburg J J (2004). Interspecific and Intraspecific Variation of δC and δN in Deposit- and Suspension-Feeding Bivalves (*Macoma balthica* and *Cerastoderma edule*): Evidence of Ontogenetic Changes in Feeding Mode of *Macoma balthica*. *Limnology and Oceanography*, **49** (2), 408–14.

Teck S J, Halpern B S, Kappel C V, Micheli F, Selkoe K A, Crain C M, Martone R, Shearer C, Arvai J, Fischhoff B, Murray G, Neslo R and Cooke R (2010). Using Expert Judgment to Estimate Marine Ecosystem Vulnerability in the California Current. *Ecological Applications*, **20**, 1402–16.

Thrush S F, Halliday J, Hewitt J E and Lohrer A M (2008). Cumulative Degradation in Estuaries: The Effects of Habitat, Loss Fragmentation and Community Homogenization on Resilience. *Ecological Applications*, **18**, 12–21.

Thrush S F, Hewitt J E, Lohrer A and Chiaroni L D (2013). When Small Changes Matter: The Role of Cross-Scale Interactions between Habitat and Ecological Connectivity in Recovery. *Ecological Applications*, **23**, 226–38.

Thrush S F and Whitlatch R B (2001). Recovery Dynamics in Benthic Communities: Balancing Detail with Simplification. In: Reise K, ed. *Ecological Comparisons of Sedimentary Shores*, Springer-Verlag, Berlin, pp. 297–316.

Thrush S F, Whitlatch R B, Pridmore R D, Hewitt J E, Cummings V J and Maskery M (1996). Scale-Dependent Recolonization: The Role of Sediment Stability in a Dynamic Sandflat Habitat. *Ecology*, **77**, 2472–87.

Van Blaricom G R (1982). Experimental Analysis of Structural Regulation in a Marine Sand Community Exposed to Oceanic Swell. *Ecological Monographs*, **52**, 283–305.

Van Colen C, Montserrat F, Vincx M, Herman P M J, Ysebaert T and Degraer S (2008). Macrobenthic Recovery from Hypoxia in an Estuarine Tidal Mudflat. *Marine Ecology Progress Series*, **372**, 31–42.

Van Colen C, Montserrat F, Vincx M, Herman P M J, Ysebaert T and Degraer S (2010). Macrobenthos Recruitment Success in a Tidal Flat: Feeding Trait Dependent Effects of Disturbance History. *Journal of Experimental Marine Biology and Ecology*, **385** (1–2), 79–84.

Volkenborn N and Reise K (2006). Lugworm Exclusion Experiment: Responses by Deposit Feeding Worms to Biogenic Habitat Transformations. *Journal of Experimental Marine Biology and Ecology*, **330**, 169–79.

Wootton J T (1998). Effects of Disturbance on Species Diversity: A Multitrophic Perspective. *American Naturalist*, **152** (6), 803–25.

PART II

Designing research

CHAPTER 4

Design and the philosophy of sampling

4.1 Introduction

Empirical research is a heuristic process, it's learning by doing. Learn from your success and others' failures. Expect surprises and be adaptable in the way you design your studies so they fit with the nature of your study system.

In particular, it is important to recognise how our view changes as we shift our focus on the world. Think of the view of the intertidal flats you have as you walk across the flats, or as your aircraft takes off from the adjacent runway and climbs into the sky. Walking, you can focus on the small animals on the sediment surface or the traces they leave; as you climb into the air you will see changes in the wetness of the sediment, patches of seagrass or shellfish and the relationships between them; higher again, the sandflat will be seen in the context of the whole harbour and adjacent land and open coast, revealing relationships between different sandflats, different land uses and exposure to the open ocean. How much we can see is the scale of the study and drives what questions it might be able to answer (Figure 4.1).

The world of the soft-sediment organisms is a partially fluid medium as sediment moves with both waves and currents, food moves with the sediment and water, and many organisms actively or passively disperse in the water, either on a tidally driven or daily basis or at certain stages in their lives. Soft sediments are less dominated by competition than rocky substrate ecosystems, because of this mobility and the 3-dimensional nature of the habitat that allows organisms to space themselves

out. Many organisms are cryptic, living their lives within the sediment, sometimes displaying signs of their presence on the sediment surface through feeding tracks, holes or mounds, still others leaving no sign at all of their presence let alone abundance. Larger epifauna and flora live their lives on the sediment surface, frequently creating a biogenic habitat, within which others live their lives.

Soft-sediment ecologists face the same challenges that other areas of ecology face: what do we want to know; can we study it at the scale that we think things are happening on; and can we create generalities that will allow us to understand whether the same thing will happen elsewhere and if not why not? However, the cryptic nature of many of the organisms, the fluid nature of the environment and the ability of many organisms to alter the environment and facilitate other species, create a unique situation for research. In this chapter we discuss the unique set of problems for sampling that this creates and the philosophy required to design studies that are strongly linked to the scale at which organisms experience their environment and biological interactions occur. For in-depth coverage of these issues we suggest the following: Eberhardt and Thomas (1991), Hewitt et al. (2007) and Quinn and Keough (2002).

4.2 Philosophical considerations

4.2.1 World views

How you view your system influences study design. Do you see the system as a homogeneous

Ecology of Coastal Marine Sediments: Form, Function, and Change in the Anthropocene. Simon F. Thrush, Judi E. Hewitt, Conrad A. Pilditch and Alf Norkko, Oxford University Press (2021). © Simon Thrush, Judi Hewitt, Conrad Pilditch, and Alf Norkko.
DOI: 10.1093/oso/9780198804765.003.0004

Figure 4.1 Scales of observation (mm–km) influence the way we think about processes and how plants and animals interact with their environment, with heterogeneity occurring at multiple scales: from features associated with individual tubes, to tube mats, simple to complex habitats, seagrass patches and gravel patches in sandflats. (Photo credits Simon Thrush, Roman Zajac and Jenny Hillman.)

landscape on which you impose your study akin to how you might set up an experiment in a laboratory? Or do you have a heterogeneous landscape in which you will nest your experiment or sampling (this means you will need to sample extra variables to account for the heterogeneity but you are also more likely to be able to create more real-world generalities)? Is the system complicated and can it only be studied by reducing the complications to a single process that can be isolated, or is it a complex system where emergent processes vary with the scale of the study? Is a community driven by strong interactions between a few species or do weak interactions and indirect effects also occur? Can cause and effect only be delivered by manipulations or can a variety of lines of evidence contribute to attributing cause and effect? You might believe that the important heterogeneity is spatial and be prepared to ignore the temporal variability, or vice versa. All these viewpoints and questions have implications for how a study is designed and indeed the questions it is expected to answer.

It is likely that the relevance of these viewpoints will vary from study system to study system but designing studies around some of them will provide a design that is more robust if another viewpoint proves to hold true in your system. For example, designing a study to overlie a homogeneous landscape and finding that the landscape was not homogeneous results in high variability of response and (usually) a 'no effect detected' result. Whereas, designing a study to overlie a heterogeneous landscape and finding that the landscape was actually homogeneous results in low variability of response and a strong ability to detect effects. So please do think carefully about what your question is and how best to answer it!

4.2.2 Rigour and generality

We all agree on the need for rigorous design and robust results (e.g. Eberhardt & Thomas, 1991; Legendre et al., 2004; Underwood et al., 2000). However, there are two aspects to this: the design allows for the question (and only that question) to be answered; or the design allows for an improved understanding of the processes that underpin ecological relationships. The latter requires an ability to generalise that is rarely answered by purely statistical considerations. Answers from random sampling (see Figure 4.2A) only apply within the population sampled and allow for only the original question to be answered. Careful location of samples along gradients and measurement of variables that represent either confounding or explanatory factors can convert spatial and temporal variability

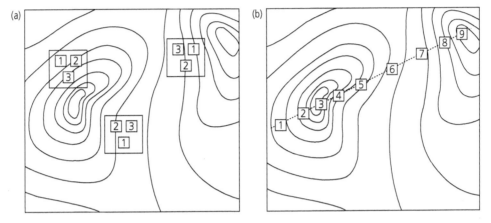

Figure 4.2 Different types of study design: (A) randomised block, (B) gradient. Ellis et al. (2000). The contour lines represent variables in a key variables (e.g. density or richness). Reprinted from *Journal of Experimental Marine Biology and Ecology*, **366** (1–2), Ellis J, Schneider D C, Spatial and Temporal Scaling in Benthic Ecology, 92–8., Copyright (2008), with permission from Elsevier.

Box 4.1 Basic design steps

Once you have your question (and your viewpoint), there are a set of design steps that can be followed (summarised in Figure 4.3 and laid out in full in the text in this box). The results at each step depend on answers to the questions posed within the following sections on scale, mobility and context dependency, but the actual steps do not.

1. Define the representativeness of your study.
2. Can the question best be answered by:
   a. Experiments within confined systems, for example laboratories and mesocosms. Apart from requiring strong consideration of scale dependence (see section 4.3), there are also artefacts that can occur (see Carpenter, 1996).
   b. Field manipulation. Again, this requires a strong consideration of scale-dependence (see section 4.3), although many of the artefacts that affect confined system experiments can be avoided. However,

exclusion and caging experiments have well-known artefacts associated with them (see Thrush, 1999). Importantly, it is not always possible to completely isolate your process of interest (Thrush & Lohrer, 2012) and some processes cannot be manipulated on the scale at which a field manipulation can be done.
   c. Field survey. Generally, these create no artefacts and can be conducted across scales. However, when trying to understand species interactions and sample temporally dynamic environmental variables (e.g. turbidity), it can be difficult to accurately position samples.
   d. Combinations of the above. For example, experiments can be nested within measured larger-scale patterns (Menge et al., 1994; Thrush et al., 2017), broad-scale environmental gradients (Keddy, 1991) or temporal cycles. Companion experimental manipulations and

continued

Box 4.1 *Continued*

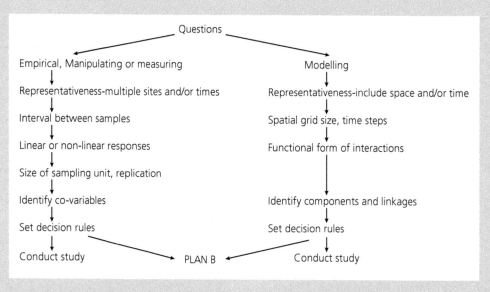

Figure 4.3 Summary of steps to follow in designing a study.

surveys may be carried out over the same extent, as Bell et al. (1995) demonstrated when they matched results of a multisite experiment on drift algae accumulation in seagrass beds with a survey in natural seagrass beds. A series of small studies can be conducted within an integrated framework to build up information (Hewitt et al., 2007).

   e. Iteration between aspects of the above and modelling.

3. Does the question require:
   a. A single site
   b. Multiple sites
   c. Hierarchical sampling?

4. Decide on sampling intervals in space or time. If the study is replicated in space, the distance you leave between survey samples or experimental plots will be driven by the probability of spatial scale dependence. Similarly, if the study is replicated in time the time interval between samples will be driven by the probability of temporal scale dependence.

5. Decide whether you expect a linear or non-linear response. If you expect the response to be non-linear, whether it has a threshold in it or is unimodal, and you want to use experimental manipulations, then there are implications for the design. Few levels and high within-level replication will not be effective; instead, multiple levels (even with low replication) will help determine

where a threshold occurs or the range in which changes in direction happen (Figure 4.4). Analysis should focus on the manipulation levels as a continuous factor. Similarly, for a survey, a design of few replicates at numerous positions along a gradient will be most appropriate.

6. Define the sampling units to be used. This is a complex procedure, requiring consideration of the size of sampling units (see Chapter 5), size of treatment plots/sites, numbers of treatment levels (or sites), the magnitude you want to detect and, of course, the cost! It also includes whether transects or grids should be used and whether samples should be collected below the level of variability of interest and pooled either to average over small-scale variability or to remove problems associated with pseudo-replication.

7. Identify co-variables, the scale at which they will be collected and the degree of replication required. We strongly recommend that, regardless of the type of study, co-variables should be collected. Even in a laboratory experiment based on manipulated treatment levels, information on the actual values of the manipulations (e.g. actual pH concentrations that each replicate of each treatment was exposed to) may reveal important results. Demonstrating that there is no spatial variation in possible confounding factors (e.g. light and temperature) will make your study more rigorous.

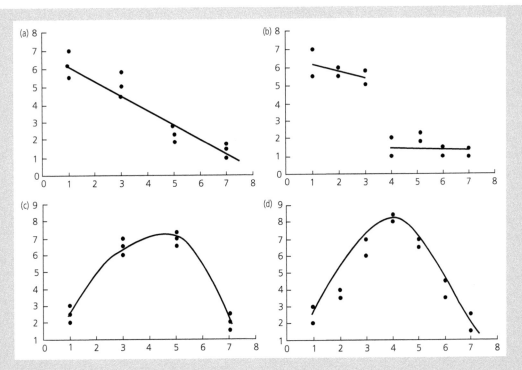

Figure 4.4 Collecting fewer replicate samples and having more treatment levels or sites along a gradient allows the apparently linear trend in Panel A to be resolved into a distinct threshold response (Panel B) and the magnitude and location of the maxima in panel D to be more precisely resolved than in (Panel C).

8. Set your decision rules. This is particularly important for experimental manipulations that are sampled over time and follows on partly from step 2. If you have conducted a manipulation, what time interval is required before the treatment reflects the process you are interested in rather than the disturbance you have caused while doing the manipulation? Often the answer to this is to collect more frequently than you think you need and use some indicator to tell you whether you need to process a sample time

or not. If you are looking at disturbance and recovery and the controls and experimental plots do not converge, will you keep sampling forever? Frequently the best you can do is to stop sampling after some length of time and analyse relative recovery as differences driven by measured co-variables (Thrush et al., 1996).

9. Always have a plan B—things hardly ever work out the way you expect.

from noise into information (Ellis & Schneider, 1997; Figure 4.2B). Understanding why responses vary between locations allows the results to be extended from a specific location (Cottenie & De Meester, 2003) to more general situations (Belovsky et al., 2004; Bestion et al., 2019; Thrush et al., 2000) (Box 4.1).

4.2.3 Attributing cause and effect

In soft-sediment marine studies there has been a history of conducting manipulative experiments under the assumption that they confer causality, whereas surveys are seen as having low causality due to the potential for spurious correlations. However, assigning causality can be achieved through multiple lines of evidence, of which timing (a result immediately follows a specific action) is only one (Hill, 1965; Peters, 1991; Rigler, 1982). Many of these lines of evidence are routinely used in other scientific areas. Plausibility is used in physics and oceanography, to make inferences where there is a known theory or accepted mechanism that supports the results. Medicine, especially epidemiology, often uses both plausibility and gradients in the strength of effect and consistency among studies (Fox, 1991; Susser, 1986). Finally, there is analogy (i.e. a similar cause has been demonstrated to lead to a similar effect).

4.3 Scale

Scale has profound implications for how ecological studies are conducted and their results are interpreted (Dayton, 1984). Different processes operate across different scales and what we observe depends on the size of the window we use to observe the system. However, understanding the effects of scale and incorporating them in study designs is particularly important if we are to get the most from our studies and learn why things vary from place to place (Figure 4.5). Ecological systems are heterogeneous and this is a crucial element in what makes them both interesting and functional (Legendre, 1993).

Three aspects of scale are important for all study designs and they operate in both space and time. (1) In a spatial context, the grain of the study is easily understood for experiments—it is the size of the experiment plot. However, even for surveys, when we take samples to represent a site we have some thought in our mind about the size of the site, for example we may take three replicates within a 5-m² area. Similarly, in a temporal context, when we take any sample, we have an idea about whether this represents an instantaneous point or whether the sample is a product of a certain length of time before the sampling. For example, the length of seagrass shoot may integrate over the previous month's temperature and light. (2) Lag is very simply the distance between sites or times sampled. (3) Extent is

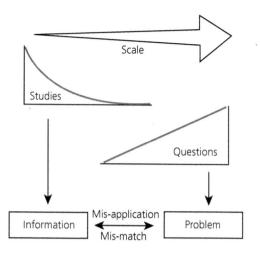

Figure 4.5 Probably the most important design question is whether the question(s) the study is to answer can be answered by collecting information at the scale of the study. Adapted from Thrush et al. (1999), Figure 3. Adapted by permission from Springer Nature: Springer, *Biogeochemical Cycling and Sediment Ecology* by Gray, Ambrose Jr and Szaniawska. Copyright of Springer Science + Business Media Dordrecht (1999).

the size or duration of time covered by the study and thus the extent of study inference.

The importance of resolution (how closely we 'look' at our samples) is the fourth aspect of scale that depends on the study topic, and is more related to data collection method, so will be discussed mainly in Chapter 5. However, most studies decide on a resolution based on the question, for example, what is the size range of organisms that the question relies on? Are you sampling for bacteria, microphytes, meiofauna, macrofauna, large macroalgae etc.?

4.3.1 Theory of scale

Four main categories of theory have been developed to link patterns observed at different spatial scales to processes: patch dynamics; hierarchy theory; gradient analysis; and multiscale theory. The earliest attempt to formally integrate the effects of processes operating on different scales was hierarchy theory (Allen & Starr, 1982), where local ecological relationships were considered to operate within a context set by environmental variables

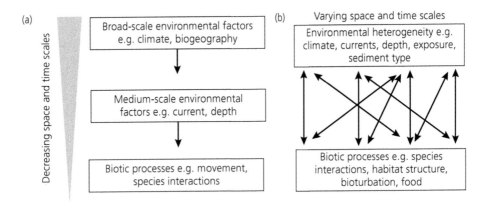

Figure 4.6 Conceptual differences between (A) hierarchy and (B) multi-scale theory.

(Kotliar & Wiens, 1990; Milne, 1991) (Figure 4.6). In many areas of ecology, but possibly especially in soft-sediment ecology, the potential for cross-scale interactions as described above was quickly recognised. Multiscale theory was then developed, where processes could operate and interact over spatial and temporal scales, including the ability of local biotic conditions to affect broader-scale processes. The ability to utilise the concept of cross-scale interactions in soft-sediment systems led to a better ability to understand context dependencies and create strong generalities from a previously confusing set of results (Soranno et al., 2014; Thrush et al., 2000, 2013).

Soft-sediment ecologists have been generally slow to utilise landscape ecology, meta-population and meta-community concepts, probably driven by the view that marine systems are open due to: their fluid nature and the mobility of many organisms (see section 4.4); and the difficulties in viewing and sampling patches. However, the ability of soft sediments to create self-organised patchiness (Rietkerk et al., 2004) and the role that meta-community theory may play in explaining community assembly processes (Valanko et al., 2015) (see Chapters 6 and 8) are increasingly being recognised.

Most of the problems caused by scale and solutions to them have been discussed in the general ecological literature for many years (see Hewitt et al., 2007 and references therein). Despite this, many marine studies have not taken up this challenge (Ellis & Schneider, 2008) and designs incorporating scale are rare. This is somewhat understandable as including scale as a factor in study design can be expensive, especially for factorial experimental manipulations. However, many studies do not state that results are likely to be scale-dependent or affected by scale mismatches, nor why the scale of the study was chosen. For example, increasingly species distribution models use environmental data generated from large-scale models, without discussing the match between the time and space scales of the modelled variables, the species data and the species' response to the environment (Hui et al., 2010). Similarly, Snelgrove et al. (2014) point out that simplified small-scale experiments on biodiversity–ecosystem functioning do not match the scales at which biodiversity loss generally occurs.

4.3.2 Scale-related design questions

The complexities that arise from the characteristics of soft-sediment systems result in a number of major considerations to study design (summarised in Figure 4.7 and discussed below) and require a flexible approach to analysis (Box 4.2).

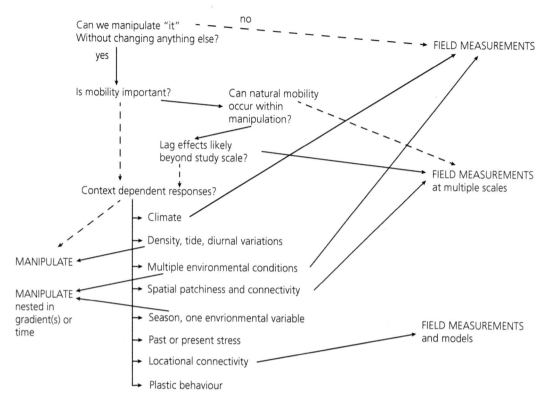

Figure 4.7 Summary of scale- and mobility-related questions that help select the type of study. Solid lines indicate 'Yes'. Dotted lines indicate 'No'. Once appropriate field data has been gathered, manipulative experiments nested into gradients can provide generality (Hewitt et al., 2007).

Box 4.2 Major considerations to study design

Specific factors that drive study design and analysis are whether:

1. The major response to a studied process occurs at scales larger than can be manipulated, or smaller than measurements can be made at. Experimental manipulations result in incorrect assumptions and correlative studies (surveys) conducted across multiple scales are needed to test predictions.

2. The manipulation creates artefacts which cannot adequately be controlled for. The effect of some artefacts can be removed by creating a control that has the same setup actions taken on it. For example, a control for an experiment with removal of large individuals by raking may simply rake the surface of the control plot. Associated with this question is whether adequate controls can be located.

3. The scale of the study is small relative to the spatial or temporal variability of the process. Samples from experimental manipulations will record high variability and 'no effects'. For example, predator–prey dynamics studied at spatial scales smaller than those of predator or prey aggregations for short time periods. Studies incorporating a variety of spatial and temporal scales, and analysed with scale as a predictor, are necessary.

4. Lags occur between a measured response and the timing of changes in environmental or biotic variables. Spatial and temporal scale need to be included in the study and lag effects need to be analysed for.

5. There are other processes of near equal importance to the process of interest that operate at similar scales and cannot be easily isolated. Applying a reductionist approach of considering that a single process can be studied will result in confusing or incorrect interpretations. Studies based on manipulative experiments and surveys need to be conducted and analysed by integrating with models.

6. Hierarchical scale theory dominates and the intensity or direction of the process or response, or indeed whether it occurs at all, depends on another, broader-scale, process (e.g. dispersal controlled by aerial or hydrodynamic processes; Thrush et al., 2000). Manipulative experiments randomly placed are likely to generate 'no effect'. Instead, manipulations need to be well placed along measured gradients in the broad-scale process. The measures of the gradient and co-variables that are associated with site differences need to be incorporated in the analysis.

4.4 Mobility

Setting aside the size of an organism, one of the obvious controllers of how species experience their environment is how mobile they are- how far and how frequently they move. In soft-sediment systems, this is frequently a combination of behavioural traits including swimming, active drifting or passive drifting and how mobile the sediment is. Bedload movement (movement of the sediment grains bouncing across the seafloor) can result in passive movement of organisms. However, many species also are actively mobile, choosing to burrow vertically or horizontally through the sediment, move across the surface of the sediment, or even enter the water column to swim or be carried by currents.

Changing the degree of mobility throughout life stages for macrofauna is common. Whereas some species brood their young, many more disperse as planktonic larvae. However, brooded young and post-settlement juveniles may swim off or enter the water column to be passively moved by currents. Even as adults, many species will swim, burrow or move across the sediment surface. For example, the New Zealand cockle *Austrovenus stutchburyi* has a planktonic larval phase, then as a post-settlement juvenile is mainly passively moved with sediment, and finally the adults burrow small vertical distances and also plough across the sediment surface for metres on a daily basis. Conversely, although the bivalve *Macomona liliana* also has a planktonic larval phase, it is actively mobile in the water column as a post-settlement juvenile, whereas adults live deeper in the sediment than do the juveniles and are relatively sedentary (moving at depth through the sediment by as little as 10 cm on a daily basis). Habitat requirements may also change with life stage, requiring organisms to move to new areas (e.g. many bivalves move to deeper waters as adults).

Differing degrees of mobility across time scales are also common. Mobility terminology is usually temporally based, with frequent (e.g. daily or tidal) movement sometimes defined as the home range, less frequent movements being dispersal and long-term movements being migration. These temporally driven aspects of mobility link to the spatial scales at which an organism experiences its environment and exploits resources. This can be ecologically important in allowing mobile organisms to exploit: temporally and spatially variable resource patches (e.g. amphipods and cumaceans using food resources that can accumulate in ray pits); or different habitats at different life stages (e.g. shellfish that, at different life stages, use different estuarine habitats with different food resources and risks of predation). Mobility can be important in moving organisms away from unfavourable conditions and recolonising disturbed areas.

Inter-individual heterogeneity due to past experiences, physiological state, and genetic and phenotypic predispositions are increasingly being recognised as important features in dispersal (Nanninga & Berumen, 2014). All this means that hydrodynamics is often not enough to remove the spatial and temporal patterns generated by biotic factors (Barry & Dayton, 1991).

4.4.1 Mobility-related design questions

If mobility is frequently larger than the size of treatment plots or sites and the organism is not able to actively initiate and end movement (see Figure 4.7), manipulative experiments are swamped by mobility and generate a 'no effect detected'. Correlative studies with samples conducted at a variety of grains can be used to determine both the effect and the likely scale of maximum effect.

If meta-population dynamics suggest that recruitment/colonisation from outside the study extent is likely to occur and the organism is not able to select where it settles (Figure 4.7), studies should be conducted at times when recruitment/colonisation is likely to be low.

4.5 Context-dependencies

4.5.1 Environmental factors

Obviously, the characteristics of the environments in which species are found are important. Regardless, a general question that we need to ask is: is the relationship observed between the environment and species actually structured by the environment immediately surrounding the species sampled? Or are the species responding to a larger scale? Common examples are edge and legacy effects, essentially lags in space and time. More complex are cross-scale interactions, generated for example by climatic patterns of rain and wind which interact with estuarine bathymetry and the presence of particular species (e.g. mangroves, cockle beds).

Moreover, the response of organisms to environmental characteristics is dependent not just on the scale at which organisms experience their environment, but also on the scale at which other processes produce spatial and temporal heterogeneity. For example, hydrodynamics present during spawning events may determine the scale at which larvae settle. The dynamic view of seafloor communities, created by a legacy of disturbances (as described in Chapter 3), also produces spatial and temporal heterogeneity within which the response of organisms to their environment occurs.

4.5.2 Species interactions

As Chapter 2 discusses, many species modify their environment over a range of scales. Importantly these modifications are often dependent on density and spatial patterns, especially where density and spatial arrangement interact with water flow. Size, density and spacing of individuals can affect species interactions. High-density patches of mobile grazers and deposit feeders can disturb surface sediments and preclude other species from establishing (Lohrer et al., 2008). For some species, whether they act as sediment stabilisers or destabilisers depends on their density (see Chapter 2). The important point is that complex interactions can accumulate over individuals to emerge at single or multiple patch scales (van de Koppel et al., 2012), and indeed can occur at multiple scales when patchiness at one scale occurs within patchiness at larger scales. Thus, the soft-sediment seafloor is a landscape of habitats and these habitats may affect what you see.

4.5.3 Context-dependent effects on design

Context dependency limits our ability to generalise beyond the study conditions. At the same time, it provides the opportunity to understand much more about how a system works and how to predict the

impacts of changing conditions, whether dependencies are driven by climate or environmental change, removal of habitat structure or key species, or biodiversity loss.

Specific factors that drive study design and analysis are included in Figure 4.7 and expanded upon below:

- If temporal variations in mobility, species–species or species–environment interactions or intraspecific interactions occur, spatial and temporal scale need to be included in the design of the study.
- If outcomes are not only density-dependent but also dependent on spatial arrangement, spatial scale (grain and lag) needs to be included in the design and analysis of survey-type studies, which may be followed by manipulations of specific densities and spatial arrangements.
- If the response is context-dependent on past or present stressors or connectivity between patches, manipulative experiments and correlative studies need to be well placed along gradients in the other factors, and potential mechanisms need to be determined.
- If the response of the species or community to the process under study can be modified by switches in behaviour to take advantage of differing conditions, mechanistic studies need to be conducted across a range of environmental drivers, biotic habitats and community types that are most likely to result in behaviour change. Analyses will need to include as co-variables measures of these factors.

4.6 Indirect effects

Together with context-dependent effects, indirect effects create a need to balance realism against the desire to increase the ability to assign causality by manipulation. Specific factors that drive study design and analysis are:

- If feedback loops are involved (e.g. increasing mud content decreases the abundance of suspension-feeding organisms whose suspension feeding removes small mud particles from the water column and places them on the seafloor, making it muddier). Small-scale mechanistic experiments and large-scale correlative studies can be used to provide parameters for network models.

- If indirect effects are generated. Small-scale mechanistic experiments and large-scale correlative studies that measure co-variables likely to be affected can be used to provide parameters for network models.
- If cross-scale interactions are involved. Large-scale correlative studies that include measurements made at a variety of scales are analysed using a mix of local and broader-scale predictor variables.

4.7 Close out

Ecological patterns range from millimetres to across ocean basins and from seconds to the expanse of evolutionary history, yet our ability to manipulate the processes which cause these patterns is generally confined to small spatial and temporal scales. Because of this limitation, manipulations cannot be our primary source of understanding. The cryptic nature of many species, together with a high probability that most of the organisms sampled are mobile, suggests that surveys across gradients of interest are likely to be more productive. We know that the patterns we observe result from complex interactions and that a simple cause–effect study conducted in different locations (or even the same location at different times) will often provide different results. However, there are simple design rules and analyses that will allow us to untangle these complexities and reveal underlying general patterns.

Heterogeneity and the various interactions between mobility, environmental characteristics, creation of biogenic habitats, key species and the role of weak interspecies interactions all reinforce the central role of scale in study design. They also lead to the conclusion that many of the observations we make for soft-sediment ecology will be context dependent. Generalities therefore need to be sought by understanding key contexts, which interactions are most important at which scales, and the paths by which indirect effects operate. Together the indication is that conducting simple cause–effect

studies will generate confusing results, but that basing studies on a mix of mechanistic and context dependencies will generate strong understandings of what happens, and why, in specific environmental and biotic contexts.

Most of what we know is based on studies over differing locations; only rarely are studies temporally replicated. Frequently we assume that the changes we observe over a spatial gradient, in for example sediment type, will be the same as those we would observe if the change in sediment occurred over time at a site (space for time surrogacy). Whether this is correct, and what factors are likely to control how correct this prediction is, are largely unknown. More studies conducted over time are necessary for us to become confident in how temporally general what we know is.

References

Allen T F H and Starr T B (1982). *Hierarchy Perspectives for Ecological Complexity*, University of Chicago Press, Chicago, IL.

Barry J P and Dayton P K (1991). Physical Heterogeneity and the Organisation of Marine Communities. In: Kolasa K and Pickett S T A, eds. *Ecological Heterogeneity*, Springer-Verlag, New York, pp. 270–320.

Bell S S, Hall M O and Robbins B D (1995). Towards a Landscape Approach in Seagrass Beds: Using Macroalgae Accumulation to Address Questions of Scale. *Oecologia*, **104**, 163–8.

Belovsky G E, Botkin D B, Crowl T A, Cummins K W, Franklin J F, Hunter Jr. M L, Joern A, Lindenmayer D B, MacMahon J A, Margules C R and Scott J M (2004). Ten Suggestions to Strengthen the Science of Ecology. *Bioscience*, **54**, 345–8.

Bestion E, Cote J, Jacob S, Winandy L and Legrand D (2019). Habitat Fragmentation Experiments on Arthropods: What to Do Next? *Current Opinion in Insect Science*, **35**, 117–22.

Carpenter S R (1996). Microcosm Experiments Have Limited Relevance for Community and Ecosystem Ecology. *Ecology*, **77** (3), 677–80.

Cottenie K and De Meester L (2003). Comment on Okansen (2001): Reconciling Okansen (2001) and Hurlebert (1984). *Oikos*, **100**, 394–6.

Dayton P K (1984). Processes Structuring Some Marine Communities: Are They General? In: Strong J D R, Simberloff D, Abele L G and Thistle A B, eds. *Ecological Communities: Conceptual Issues and the Evidence*, Princeton University Press, Princeton, NJ, pp. 181–200.

Eberhardt L L and Thomas J M (1991). Designing Environmental Field Studies. *Ecological Monographs*, **61**, 53–73.

Ellis J and Schneider D C (2008). Spatial and Temporal Scaling in Benthic Ecology. *Journal of Experimental Marine Biology and Ecology*, **366** (1), 92–8.

Ellis J I and Schneider D C (1997). Evaluation of a Gradient Sampling Design for Environmental Impact Assessment. *Environmental Monitoring and Assessment*, **48** (2), 157–72.

Ellis J I, Schneider D C and Thrush S F (2000). Detecting Anthropogenic Disturbance in an Environment with Multiple Gradients of Physical Disturbance, Manukau Harbour, New Zealand. *Hydrobiologia*, **440** (1–3), 379–91.

Fox G A (1991). Practical Causal Inference for Ecoepidemiologists. *Journal of Toxicology and Environmental Health*, **33**, 359–73.

Hewitt J E, Thrush S F, Dayton P K and Bonsdorf E (2007). The Effect of Spatial and Temporal Heterogeneity on the Design and Analysis of Empirical Studies of Scale-Dependent Systems. *American Naturalist*, **169**, 388–408.

Hill A B (1965). The Environment and Disease: Association or Causation? *Proceedings of the Royal Society of Medicine*, **58**, 295–300.

Hui C, Terblanche J S, Chown S L and McGeoch M A (2010). Parameter Landscapes Unveil the Bias in Allometric Prediction. *Methods in Ecology and Evolution*, **1**, 69–74.

Keddy P A (1991). Working with Heterogeneity: An Operator's Guide to Environmental Gradients. In: Kolasa J and Pickett S T A, eds. *Ecological Heterogeneity*, Springer-Verlag, New York, pp. 181–201.

Kotliar N B and Wiens J A (1990). Multiple Scales of Patchiness and Patch Structure: A Hierarchical Framework for the Study of Heterogeneity. *Oikos*, **59**, 253–60.

Legendre P (1993). Spatial Autocorrelation: Trouble or New Paradigm? *Ecology*, **74**, 1659–73.

Legendre P, Dale M R T, Fortin M J, Casgrain P and Gurevitch J (2004). Effects of Spatial Structures on the Results of Field Experiments. *Ecology*, **85**, 3202–14.

Lohrer A L, Chiaroni L D, Hewitt J E and Thrush S F (2008). Biogenic Disturbance Determines Invasion Success in a Subtidal Soft-Sediment System. *Ecology*, **89**, 1299–307.

Menge B A, Berlow E L, Blanchette C A, Navarrete S A and Yamada S B (1994). The Keystone Species Concept: Variations in Interaction Strength in a Rocky Intertidal Habitat. *Ecological Monographs*, **64**, 249–86.

Milne B R (1991). Heterogeneity as a Multiscale Characteristic of Landscapes. In: Kolasa J and Pickett S T A, eds. *Ecological Heterogeneity*, Springer-Verlag, New York, pp. 69–84.

Nanninga G B and Berumen M L (2014). The Role of Individual Variation in Marine Larval Dispersal. *Frontiers in Marine Science*, **1**, 71.

Peters R H (1991). *A Critique of Ecology*, Cambridge University Press, Cambridge, United Kingdom.

Quinn G P and Keough M J (2002). *Experimental Design and Data Analysis for Biologists*, Cambridge University Press, Cambridge, United Kingdom.

Rietkerk M, Dekker S C, de Ruiter P C and van de Koppel J (2004). Self-Organized Patchiness and Catastrophic Shifts in Ecosystems. *Science*, **305** (5692), 1926–9.

Rigler F H (1982). Recognition of the Possible: An Advantage of Empiricism in Ecology. *Canadian Journal of Fisheries and Aquatic Science*, **39**, 1323–31.

Snelgrove P V R, Thrush S F, Wall D H and Norkko A (2014). Real World Biodiversity–Ecosystem Functioning: A Seafloor Perspective. *Trends in Ecology & Evolution*, **29** (7), 398–405.

Soranno P A, Cheruvelil K S, Bissell E G, Bremigan M T, Downing J A, Fergus C E, Filstrup C T, Henry E N, Lottig N R, Stanley E H, Stow C A, Tan P-N, Wagner T and Webster K E (2014). Cross-Scale Interactions: Quantifying Multi-Scaled Cause–Effect Relationships in Macrosystems. *Frontiers in Ecology and the Environment*, **12** (1), 65–73.

Susser M (1986). Rules of Inference in Epidemiology. *Regulatory Toxicology and Pharmacology*, **6**, 116–28.

Thrush S F (1999). Complex Role of Predators in Structuring Soft-Sediment Macrobenthic Communities: Implications of Changes in Spatial Scale for Experimental Studies. *Australian Journal of Ecology*, **24** (4), 344–54.

Thrush S F, Hewitt J E, Cummings V J, Green M O, Funnell G A and Wilkinson M R (2000). The Generality of Field Experiments: Interactions between Local and Broad-Scale Processes. *Ecology*, **81** (2), 399–415.

Thrush S F, Hewitt J E, Kraan C, Lohrer A M, Pilditch C A and Douglas E (2017). Changes in the Location of Biodiversity–Ecosystem Function Hot Spots across the Seafloor Landscape with Increasing Sediment Nutrient Loading. *Proceedings of the Royal Society B*, **284**, 20162861.

Thrush S F, Hewitt J E, Lohrer A and Chiaroni L D (2013). When Small Changes Matter: The Role of Cross-Scale Interactions between Habitat and Ecological Connectivity in Recovery. *Ecological Applications*, **23**, 226–38.

Thrush S F, Lawrie S M, Hewitt J E and Cummings V J (1999). The Problem of Scale: Uncertainties and Implications for Soft-Bottom Marine Communities and the Assessment of Human Impacts. In: Gray J S, Ambrose W and Szaniawska A, eds. *Bieogeochemical Cycling and Sediment Ecology*, Kluwer, Dordrecht, Netherlands, pp. 195–210.

Thrush S F and Lohrer A M (2012). Why Bother Going Outside: The Role of Observational Studies in Understanding Biodiversity–Ecosystem Function

Relationships. In: Paterson D M, Solan M and Aspenal R, eds. *Marine Biodiversity and Ecosystem Functioning: Frameworks, Methodologies, and Integration*, Oxford University Press, Oxford, pp. 198–212.

Thrush S F, Whitlatch R B, Pridmore R D, Hewitt J E, Cummings V J and Maskery M (1996). Scale-Dependent Recolonization: The Role of Sediment Stability in a Dynamic Sandflat Habitat. *Ecology*, **77**, 2472–87.

Underwood A J, Chapman M G and Connell S D (2000). Observations in Ecology: You Can't Make Progress on Processes without Understanding the Patterns. *Journal of Experimental Marine Biology and Ecology*, **250**, 97–115.

Valanko S, Heino J, Westerbom M, Viitasalo M and Norkko A (2015). Complex Metacommunity Structure for Benthic Invertebrates in a Low-Diversity Coastal System. *Ecology and Evolution*, **5** (22), 5203–15.

van de Koppel J, Bouma T J and Herman P (2012). The Influence of Local- and Landscape-Scale Processes on Spatial Self-Organization in Estuarine Ecosystems. *Journal of Experimental Ecology*, **215**, 962–7.

CHAPTER 5

Data collection methods and statistical analyses

5.1 Introduction

Central to any study are the methods used to collect the data and the analyses used to elucidate patterns. In any field of ecology, there are multiple ways to collect data; there is a particularly wide range of methods available to soft-sediment ecologists (see Figure 5.1). The need for this large range of methods is driven by five main factors: the range of size classes of organisms; the cryptic nature of many organisms; within-sediment processes; water column processes; and their interactions. Similarly, analyses that can elucidate patterns and processes need to be able to be flexible and cope with many variables, usually at different scales, often with very different ranges of numbers and high numbers of zeros in community data matrices.

The difficulties in collecting and analysing such data, linked to increasing technologies, have driven a rapid development and diversification of data collection and analysis methods, such that it is impossible to cover any of them in any detail in a single chapter. Throughout we reference review papers and standard books, but for general data collection texts see Boudreau and Jørgensen (2001), Bianchi and Canuel (2011) and Eleftheriou (2013); and for statistical analyses see Borcard et al. (2018), Grace and Irvine (2020), Quinn and Keough (2002), Zar (1984) and Zuur et al. (2010). However, the focus of the section on data collection methods is their link to study design in terms of sample unit size (grain) and their ability to provide information at the resolution wanted for different studies. We briefly mention the link between scale, heterogeneity, replication and cost. In the section on statistical analyses we discuss the categories of analyses available and focus on some important questions to answer while selecting an analysis type.

5.2 Data collection methods

All data collection methods have their own strengths and weaknesses and there are some basic questions, generally related to scale, that need to be answered to determine whether the collection method is appropriate for the question and how many replicates are needed. This section discusses some important aspects of the scale of data collection methods, from a study perspective. Three of the four aspects of scale referred to in section 4.3 are also important for considering data collection methods, although their definition has changed slightly.

5.2.1 What area or duration are data being collected over?

The grain is the size of the area or duration of time over which an individual replicate is collected. With many methods this is easy to answer. You take a core or an image of a quadrat and its size-well, is its size. You take a porewater sample from the sediment or from within a benthic incubation chamber and the duration of sampling is how long it takes to get a sample to measure what you want to measure.

However, for many remote methods this is no longer the case. For example, the area over which data are collected by a single ping of a multi-beam acoustic device is dependent on the water depth.

Ecology of Coastal Marine Sediments: Form, Function, and Change in the Anthropocene. Simon F. Thrush, Judi E. Hewitt, Conrad A. Pilditch and Alf Norkko, Oxford University Press (2021). © Simon Thrush, Judi Hewitt, Conrad Pilditch, and Alf Norkko.
DOI: 10.1093/oso/9780198804765.003.0005

Figure 5.1 Sampling soft sediments for pattern and process studies. Sampling devices with a set grain: (a) microelectrode; (b) eDNA sediment sample; (c) macrofaunal core nested in a video transect; (d) grab sample; (e) sampling benthic chamber; (f) video transect; (g) EROMES to measure erosion rates in a core; (h) field-deployable annular flume to measure erosion rate Sampling devices with a variable grain; (i) fish trap; (j) multibeam image of reef and sediments and (k) oxygen electrode and ADV to measure eddy correlation oxygen flux. Images courtesy of Simon Thrush, Dana Clark, Greig Funnell, Rod Budd, Jenny Hillman, Peter (Chazz) Marriott, Conrad Pilditch.

Although this can be calculated when depth is relatively constant, it becomes very difficult when measurements are being made over a sloping seafloor (Eleftheriou, 2013). For eDNA analysed from sediment samples, we do not fully appreciate the subtleties of whether the species identified is actually in the core sample, or has merely passed through it in some period of time. If the eDNA is analysed from a water sample, the data may come from nearby or a long way away; the distance that the e-DNA has travelled will affect both spatial and temporal grain. A recent survey of eDNA studies (Mathieu et al., 2020) identified gaps in the scales over which research has been conducted. The survey concluded that more long-term and broad-scale eDNA studies are necessary to determine their usefulness for monitoring effects of environmental change.

For other methods, the area (and duration) over which data are collected is the subject of study. The Aquatic Eddy Covariance (AEC) technique has emerged as an important method to quantify *in situ* seafloor metabolism and primary production, measuring oxygen fluxes from a seafloor area of approximately 80 m^2 within a 5-m upstream distance of the instrument (Rodil et al., 2019). These distances are likely to be current-dependent.

Even some direct measurements can cause problems. There are multiple devices available to measure sediment stability/erosive ability, varying from fluid shearing from horizontal currents/bed shear stresses measured in flumes to vertical water jets (EROMES and CSM) (Widdows et al., 2007). These methods not only exert different forces but measure over different areas, resulting in a need to increase replication if comparisons are to be made with studies using a different device.

These previous examples are all considering grain in a spatial sense. But we have to consider the temporal grain of our data collection method as well. Again, for some methods it is simple, for example, selecting the frequency with which pressure sensors collect readings. For others, the duration over which data are collected is driven by the need to be able to gain a measurable response, for example, when collecting oxygen data from chambers to be used to calculate efflux and influx, using gels which allow adsorption of specific chemicals onto their surface, or using the decay rates of isotopes to resolve temporal differences in sediment layers (e.g. lead or caesium; Drexler et al., 2018).

However, occasionally the temporal grain is left unspecified, although hopefully it is being thought of. For example, when a macrofaunal core is taken, we rarely consider that individuals found in the core are an instantaneous measure. Rather, we think of their presence in two temporal ways, dependent on what we are studying. If we are investigating small-scale correlations with the environment, we might consider that a core represents what will be found over hours, whereas if we are monitoring for environmental change we will generally consider that they integrate over days to months (but see Chapter 4). In many cases there can be a match, or mismatch, in the time or space scales between the object of study (e.g. benthic macrofauna) and its drivers (some environmental variables of interest). Hence it is important to think about how variables match, even though they operate over different time and space scales, to make the linking of them relevant in the first place (see also section 5.3.3).

It's also important to consider whether you are measuring the actual variable or a proxy—and if the latter, does the scale the proxy is being collected at match that of the actual variable?

For example, measuring primary production can be done using a variety of methods. The rate of carbon assimilation in plant tissues, whole plants or plankton samples can be quantified by a number of tracer techniques using, for example, radioactive isotopes of C_{14}, or stable isotopes of C_{12} and C_{13}. Most commonly, in benthic ecology, primary production is calculated as the sum of the rate of change over a fixed time period in oxygen concentration in bottles or chambers. Biomass increase and increases in shoot length of seagrasses or other macrophytes are also commonly used methods that serve as proxies for assessing primary production. All these are measures from different temporal scales. At broader spatial scales, the AEC method (discussed above) allows for the quantification of primary production and respiration at the scale of habitats (10–100 m²). Similarly, secondary production is estimated by multiple methods (see Gray & Elliott, 2009 and http://www.thomas-brey.de/science/virtualhandbook/). Brey's online resource details the assumptions, conversion factors and various models for calculating growth, secondary production and productivity.

5.2.2 What is the data resolution?

The resolution of what is being measured by the sampler is another important consideration. With traditional methods for collecting data on plants and animals, this is fairly simple to estimate. The mesh size of the sieve used, or the ability to visually identify plants in quadrats, gives the smallest size of organism (and thus the resolution) that can be collected. For chemicals, the detection limit is the resolution.

But there are many other types of resolution, for example taxonomy and chemical compounds, habitat and functional grouping, abundance, or temporal resolution. Below we briefly provide examples of each of these.

Taxonomic resolution, done by traditional methods, is chosen by the analyst, although this may be restricted by the availability of identification keys. Lack of keys, for example, usually means that most macrofaunal communities contain multiple levels of taxonomic resolution, even within a phylum. For example, some families of polychaetes may be identified to species level, whereas others may only be able to be identified to family level. Although extraction of DNA and RNA and species barcoding may be able to increase our ability to define the number of species in a sample, identifying which species they are is still limited by the number of species that are barcoded.

Similarly, a study of nitrogen in porewaters may measure individual nitrogen compounds, or may only measure total nitrogen.

Habitat categorisation is a resolution applied in many surveys, and again is usually chosen by the analyst. For video transects or images, the actual categories used are generally chosen after an initial inspection of the data. For acoustic data, categorisation is usually achieved by using statistics to identify homogeneous areas. This has resulted in an often less than satisfactory ability to create habitat or community groups, despite much work in the mid-1990s recommending the use of acoustics to map benthic biological features (e.g. Magorrian et al., 1995; Schwinghamer et al., 1996). Unfortunately, despite the length of time since acoustics were first used, descriptions still generally are discriminating between rocky, biotic reef structures and 'other' soft sediment. Those that attempt a greater resolution generally find a wide range of communities within a habitat in a location (e.g. Kostylev et al., 2001; Zajac et al., 2003) and between locations (Bowden & Hewitt, 2011).

Functional categorisation is in some cases linked to habitat categorisation, for example, per cent cover of large multibranched macroalgae. Its use in mapping has increased with the desire to quickly map areas that differ in ecosystem function, in sensitivity to human activities and in the production of ecosystem services. However, DNA and RNA that initially allowed only description of bacterial communities (Dowle et al., 2015) are now being used to directly estimate bacterial function (Wemheuer et al., 2020).

Another aspect of resolution is the magnitude of what is being recorded. Is abundance (or per cent cover), biomass/size or merely presence/absence recorded? This is an important decision as extra information on processes and responses to stressors can be derived from quantitative data. This is still a major drawback to the use of eDNA which can provide relative abundance and biomass estimates (but see Fonseca, 2018), but not size information. Conversely, it can provide cross-size class community composition (e.g. bacteria through to macrofauna) (Lejzerowicz et al., 2015) and methods are developing rapidly, which may provide promise for the future.

Finally, there is a temporal resolution to be considered, especially in samples of organisms. Generally, we only want to include organisms that are alive at the time of sampling; this can be determined by the use of organic stains in sediment samples that are visually counted or by visual assessment of epibenthic animals.

5.2.3 How far apart should replicates be placed?

Lag is the distance between sample units or time between samplings, and the lag selected is driven mainly by the need to collect independent measurements. This independence is needed for replicates that are to be used in statistical analysis. Lags are therefore a balance between collecting replicates that are sufficiently similar to be useful but that are not spatially or temporally correlated and result from the spatial heterogeneity of the area and what is known about temporal variation (see section 8.3.1 on fast and slow processes). There is an exception to this: if the study is investigating spatial or temporal patterns (see section 5.3.1), in which lags are set to be the smallest distance or time difference that the study is interested in. In this case there may actually be no distance between samples.

However, there is another way that we can consider lags, especially if we are collecting data for correlation or regression analysis: that is, edge effects. For example, the relationship between the density of a 3-dimensional structuring organism and horizontal water flow above depends on how far from the start of the dense patch the data are collected. The relationship between seagrass and the number of species observed in a seagrass patch depends not only on the per cent cover of the seagrass but also on how far from the edge of the seagrass patch the sample is collected.

5.2.4 Replication and cost considerations

By this stage in determining how to answer our study question(s) we have considered the number of sites and/or treatment levels (Chapter 4) and the grain, lag and resolution of our data collection methods for the variables of interest. Now we need to decide how many replicates we can or need to process. Do we spread a lot of small samples across our study site or a few big ones? To some extent that is a relative question driven by the spatial scale of the study process or the organism's interaction with the environment. For example, larger cores are needed for macrofauna than meiofauna and microphytes in order to sample sufficient numbers or even fit the larger animals into a core.

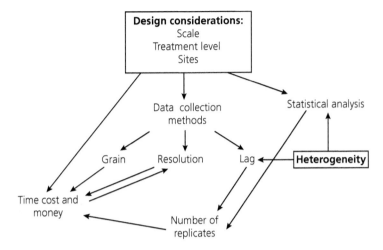

Figure 5.2 The interplay between multiple factors drives the choice of number of replicates. The primary drivers are the study question and thus its design, and the spatial and temporal heterogeneity of the study area.

But apart from that, replication is also driven by practicalities (Figure 5.2). It's at this stage that we commonly find we need to make compromises because we have neither the time nor the money to collect and process everything we want! Here it is important to have thought about the statistical analyses that are going to be used to interpret the data. Different analyses will be better able to detect patterns from your data (see Chapter 4 and section 5.3).

There are numerous methods available to determine the number of samples that are likely to be required to detect different magnitudes of effects for different study designs and analytical methods (Cohen, 1988; Edgington, 1987). But all these methods depend on some prior knowledge of the spatial and temporal heterogeneity of what you are interested in. Should you base your replication on other people's findings from a different place and time? Or instead, should you use some of your precious time and money collecting and analysing initial information, when you don't know if the variability you see now will be the same later? Desperation may lead you to just throw a dart at a dartboard! One way through this is to essentially merge the 'pilot' study into the study sampling. That is, collect more replicates, or more variables, than you can analyse. Decide on a set of 'necessary' variables and 'minimum number of replicates' that

doesn't use up all your time and preferentially analyse these before your next sampling time. Then, as your analyses proceed, put more effort where you need it.

5.3 Statistical analyses

Studies of soft-sediment ecology were initially very descriptive, with very little statistical analysis (Gray, 1981). They then developed into fairly focussed experimental analyses; now, an ever-expanding field of methods is used. The first extension was logical from an ecological perspective; multivariate analyses of community data were merely waiting for computing power to catch up with need. The second appears equally logical: the use of regression-type analyses either to allocate the importance of different drivers or to predict responses to specific environmental variables. The third was statistical in nature: the development of statistics that could cope with the non-normal, often highly skewed nature of count and other ecological data, without the loss of power and flexibility of some non-parametric methods. Since then, the types of categorical- and regression-type analyses have burgeoned, merging exploratory and predictive analyses and varying from using the data to drive the analysis to testing whether theory is matched to empirical data. Acronyms abound, from ANOVAs, to GLM, GLzM, GAMM, BRT, RF, RDA and GDM among others (see the following sections).

Regardless of the actual method chosen, there are still some important decisions to make, related to study questions, that may aid with your choice of method.

5.3.1 Making use of variability

An important question, often not really considered in advance, is: is the focus of the study on the mean response or is it important to know the range or variability of responses? Discussion around this point was initially generated by studies demonstrating that species responses can become more or less variable when the species is under stress. This led to calculations of variance being used as the response variables in ANOVAs or regressions. Over time, multivariate techniques for calculating variance have been developed and can be used in similar ways. However, we may also want to understand the range of responses, or whether the highest (or lowest) responses are controlled by some factor, e.g. body size and abundance (Blackburn et al., 1992). This was developed firstly into the concept of factor ceilings (Thomson et al., 1996; Thrush et al., 2003; see Figure 5.3) and later into the ability to produce regressions for any percentiles so that they could be compared (i.e. quantile regressions; Anderson, 2008; Cade et al., 1999, 2005).

Still focussed on variation, we need to consider how we are going to treat spatial and temporal patterns. Spatial and temporal patterns may be caused by resource or hydrodynamic patterns and their effect on the study question may be resolved by measuring those. However, especially for species abundances, they may be a result of inherent dynamics of the species, or of species–species interactions. Early indicators of the degree of aggregation in the distribution of species were indices like the variance/mean ratio (v/m) that are used to determine whether species abundances are even (v/m < 1), random (v/m = 1) or aggregated (v/m > 1). Although this type of index indicates if the spatial distributions are clumped, they do not provide information on the actual spatial pattern. Measurement of the patch structure (either tem-

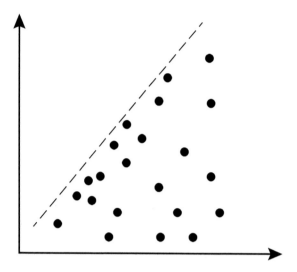

Figure 5.3 Responses to a measured factor are highly variable but cannot cross an upper limit indicated by the dashed line. The implication is that some factor(s) is forcing a ceiling on this specific relationship.

porally or spatially) may be useful to indicate the scale of the process causing the patches and thus the effect on study design (particularly replication) and interpretation. There are numerous statistical techniques for understanding and, if desired, removing these patterns. However, understanding the factors driving variance and the types of patterns it produces is important. These understandings contribute to the development of mechanistic understanding of processes. Semi-variograms, distance-decay functions, Moran's I and Geary's C indices can be used to understand the scale of spatial patchiness (Cliff & Ord, 1974). Many of these can also be used to look at temporal patterns but the use of autoregressive moving averages and differencing (ARIMA) is more common (Box & Jenkins, 1976; Chatfield, 1980). For multivariate analyses, Principal Coordinates of Neighbourhood Matrix or Moran's eigenvector mapping can be used to extract both spatial and temporal patterns (Borcard et al., 2004). Understanding the scales of these spatial and temporal patterns can be important itself (and generated the field of landscape ecology), but this understanding is also essential to measuring resilience and predicting the approach of tipping points (Chapter 8).

5.3.2 Analysing for non-linear responses

An important question raised in Chapter 4 is whether the response is expected to be linear or non-linear. Initially, non-linearity was addressed by categorical analyses in ANOVA, and regression used for linear responses. However, the desire to describe non-linear dynamics better—e.g. to understand optimum ranges or to know more precisely where thresholds occur—generated a new set of methods. Incorporation of smoothing functions is used in Generalised Additive Models (GAM; Hastie & Tibshirani, 1990; Yee, 2015). Boosted regression trees (BRT; Elith et al., 2008) essentially generate non-linear responses from the data. To answer the question of where threshold responses occur, there are numerous change-point detection methods. Generally, these change-point detection methods only examine threshold responses to a single driver, although regression trees (Breiman et al., 1984) allow direct investigation of cascading responses from many drivers. GAM, BRT, Random Forests and regression trees can only be applied to a single response variable at a time, although there is a multivariate version of the regression tree method (De'Ath, 2002) and some vector methods allow multiple GAMs (Yee, 2015). Generalised dissimilarity models allow for prediction of non-linear turn-over along environmental gradients (Ferrier et al., 2007).

5.3.3 The quest for generality

Another important question is related to how we intend to create generalities or understand context dependencies. Although Chapter 4 suggests the collection of co-variables to help with this, these are rarely measured at the same scale as the primary response measures. Information on climate variables (temperature, rainfall, wind, SOI and NAO indices) is usually available at the frequency of hours to a month. Hydrodynamic data are often collected at Hz frequencies and water quality data can be collected at frequencies from 15 mins to once a month. Therefore, if an experiment of two months' duration has been conducted, some decisions have to be made as to what values to use (Figure 5.4). Similarly, if a survey has been conducted, what is the time lag most likely to be affecting the collected data? Cross-correlations can be used to determine the most likely lag, but a more informed analysis can be gained from also including as predictor variables, various percentiles of the statistical distributions of the data rather than means. Hewitt and Norkko (2007) discovered that the effect of suspended

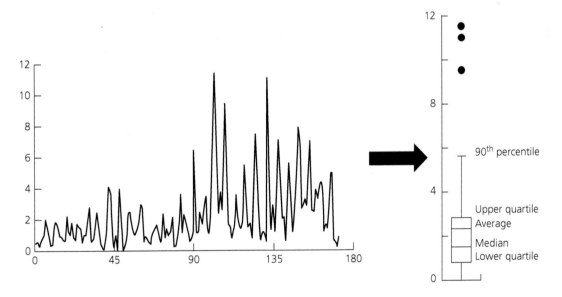

Figure 5.4 Co-variable data collected at a finer scale than the main response variable may usefully be reduced to statistical information.

sediment on juvenile bivalve mortalities was negatively correlated with the upper quartile and positively correlated with the 90th percentile. These results highlight that the effects of suspended sediments on juvenile bivalves are driven by both concentration and duration.

Similarly, co-variables collected to investigate cross-scale interactions are never all measured at the same scale. In this case, the analysis usually compares the ability to predict based only on local-scale variables, broader-scale variables and finally using variables across a mix of scales (Thrush et al., 2010).

Co-variables can also be used in meta-analyses to draw out the factors driving similarities and contrasts between studies. The term 'meta-analysis' can simply mean combing p-values and degrees of freedoms across studies into a final test of enhanced predictive power. However, the results of studies that have included the measurement of co-variables offer the ability to create a new variable from each study that reflects the results of interest (e.g. strength, magnitude and direction of the relationship between oxygen fluxes and mud content) and then analyses the effect of other variables on that result. For example, linear regression or regression tree analysis of recovery rates against current speed, inundation rates and distance to the nearest similar community can allow us to determine which of these factors are important (Thrush et al., 2008).

5.3.4 Prediction vs exploration vs theory testing

Although the line between these different tasks is often blurry, we should have in mind whether we are essentially interested in being able to predict a response or in exploring how connections and relationships play out. There are two simple results of making this decision. Firstly, when using GAM or BRT, 'wiggly' responses to a driver are often used (i.e. some variation on multiple S-shaped curves). If you're interested in exploring likely relationships, you should be asking whether you can think of any mechanistic reason why this should be the response shape before accepting it. If you're interested in prediction then what matters is whether it is the best fit and "why" comes a long second to 'it produces the best fit'. Secondly, forwards selection will pull out the best predictor first, although that prediction

may be a result of a direct effect of that variable plus an indirect effect of another variable. Backwards selection will often allow you to understand the variables involved more fully, and subsequent partialling out of effects either in ordination analysis (Borcard et al., 1992) or in path analysis (Wootton, 1994) can demonstrate whether indirect effects and modifiers are important. For example, while the bivalve *Atrina* can directly affect oxygen consumption in the overlying water column, it can also affect surrounding macrofaunal abundance and microphyte biomass which also affect oxygen consumption (Hewitt et al., 2006).

The methods discussed to date are directly allowing the data to drive the results, with null hypotheses related to no effect. However, tests of theories, or of expert opinions, are also important. Structural Equation Modelling was created around this concept (Grace & Irvine, 2020; Kline, 2011) and is increasingly being used in soft-sediment studies (e.g. Pollman et al., 2017; Thrush et al., 2012). Here, you are testing whether your data fit your conceptual model. Your null hypothesis is that there is a good fit, and thus for once you are hoping for an overall p-value >> 0.05.

5.4 Close out

Data collection methods are an area where change is happening quickly. This is one of the reasons why we have not delved deeply into individual methods, instead focussing on the general questions that need asking before you decide on any particular method. It is vital that the data collection method gives you data at the grain, resolution and lag that you need to answer your study question(s). Matching data on drivers and responses while being conscious of how they might operate over different time and space scales is imperative. You also need to be able to afford the degree of replication needed to give the statistical analyses the power to interpret patterns relevant to your questions. However, as you think about how best to answer the study questions based on the time and resources available, it is also important that you think about scale from the organisms' perspectives (see Chapter 4).

Statistical analyses readily available are also changing, although not at quite the same speed.

However, the days when ANOVA was the analysis of choice are long gone. Here we have tried to avoid repeating basic statistics guides; rather, we have provided some tips that should be useful for analysing complex patterns such as those that result from processes interacting across scales. In particular, we emphasise that trying to utilise variability, rather than dismissing it as noise, is important. Equally important is investing time in finding the statistical analysis that is most likely to allow the study question to be answered, rather than tailoring the study to a specific analysis type.

There are two areas of work that are likely to become of more importance over time. Firstly, even once the factors that are likely to drive response variability are incorporated into an analysis, there is almost bound to be some uncertainty. Increasingly, method development is looking at how this uncertainty can be displayed and understood, particularly when the information is to be used in environmental or social decision-making (Landerretche et al., 2017; Milner-Gulland & Shea, 2017; Uusitalo et al., 2018). On the more theoretical side, with increasing computing power comes the ability to move from machine learning used in many recent techniques to the use of AI (artificial intelligence) techniques. The merging of fuzzy logic, conditional probability and AI learning techniques is likely to drive this area forwards rapidly (Desjardins-Proulx et al., 2019). Use of AI based on ecological theory, learnings or expert opinions demonstrates the possibility of future analyses that can modify learnings, incorporate uncertainties and produce bounds around generalities and contexts.

References

Anderson M (2008). Animal-Sediment Relationships Re-Visited: Characterising Species' Distributions along an Environmental Gradient Using Canonical Analysis and Quantile Regression Splines. *Journal of Experimental Marine Biology and Ecology*, **366**, 16–27.

Bianchi T S and Canuel E A (2011). *Chemical Biomarkers in Aquatic Ecosystems*, Princeton University Press, Princeton, NJ.

Blackburn T M, Lawton J H and Perry J N (1992). A Method of Estimating the Slope of Upper Bounds of Plots of Body Size and Abundance in Natural Animal Assemblages. *Oikos*, **65**, 107–12.

Borcard D, Gillet F and Legendre P (2018). *Numerical Ecology with R*, 2nd ed., Springer International, Cham.

Borcard D, Legendre P, Avois-Jacquet C and Tuomisto H (2004). Dissecting the Spatial Structure of Ecological Data at Multiple Scales. *Ecology*, **85**, 1826–32.

Borcard D, Legendre P and Drapeau P (1992). Partialling out the Spatial Component of Ecological Variation. *Ecology*, **73**, 1045–55.

Boudreau B P and Jørgensen B B, eds. (2001). *The Benthic Boundary Layer: Transport Processes and Biogeochemistry*, Oxford University Press, Oxford, UK.

Bowden D A and Hewitt J E (2011). Recommendations for Surveys of Marine Benthic Biodiversity: Outcomes from the Chatham-Challenger Ocean Survey 20/20 Post-Voyage Analyses Project. *New Zealand Aquatic Environment and Biodiversity Report*, No. **91**, Ministry of Fisheries, Wellington.

Box G E P and Jenkins G P (1976). *Time Series Analysis: Forecasting and Control*, Holden Day, San Francisco, CA.

Breiman L, Friedman J, Stone C and Olshen R (1984). *Classification and Regression Trees*, CRC Press, Boca Raton, FL.

Cade B S, Noon B R and Flather C H (2005). Quantile Regression Reveals Hidden Bias and Uncertainty in Habitat Models. *Ecology*, **86** (3), 786–800.

Cade B S, Terrell J W and Schroeder R L (1999). Estimating Effects of Limiting Factors with Regression Quantiles. *Ecology*, **80**, 311–23.

Chatfield C (1980). *The Analysis of Time Series: An Introduction*, Chapman and Hall, London, UK.

Cliff A D and Ord J K (1974). *Spatial Autocorrelation*, Pion Limited, London, UK.

Cohen J (1988). *Statistical Power Analysis for the Behavioural Sciences*, Lawrence Erlbaum Associates, Hillsdale, NJ.

De'Ath G (2002). Multivariate Regression Trees: A New Technique for Modeling Species–Environment Relationships. *Ecology*, **83** (4), 1105–17.

Desjardins-Proulx P, Poisot T and Gravel D (2019). Artificial Intelligence for Ecological and Evolutionary Synthesis. *Frontiers in Ecology and Evolution*, **7**, 402.

Dowle E, Pochon X, Keeley N and Wood S A (2015). Assessing the Effects of Salmon Farming Seabed Enrichment Using Bacterial Community Diversity and High-Throughput Sequencing. *FEMS Microbiology Ecology*, **91** (8), fiv089.

Drexler J Z, Fuller C C and Archfield S (2018). The Approaching Obsolescence of 137cs Dating of Wetland Soils in North America. *Quaternary Science Reviews*, **199**, 83–96.

Edgington E S (1987). *Randomisation Tests*, Dekker, New York.

Eleftheriou A (2013). *Methods for the Study of Marine Benthos*, John Wiley & Sons, Hoboken, NJ.

Elith J, Leathwick J R and Hastie T (2008). A Working Guide to Boosted Regression Trees. *Journal of Animal Ecology*, **77** (4), 802–13.

Ferrier S, Manion G, Elith J and Richardson K (2007). Using Generalized Dissimilarity Modelling to Analyse and Predict Patterns of Beta Diversity in Regional Biodiversity Assessment. *Diversity and Distributions*, **13**, 252–64.

Fonseca V G (2018). Pitfalls in Relative Abundance Estimation Using eDNA Metabarcoding. *Molecular Ecology Resources*, **18** (5), 923–6.

Grace J and Irvine K (2020). Scientist's Guide to Developing Explanatory Statistical Models Using Causal Analysis Principles. *Ecology*, **101**, e02962.

Gray J S (1981). *The Ecology of Marine Sediments*, Cambridge University Press, Cambridge, UK.

Gray J S and Elliott M (2009). *Ecology of Marine Sediments: From Science to Management*, Oxford University Press, Oxford, UK.

Hastie T and Tibshirani R J (1990). *Generalized Additive Models*, Chapman and Hall, London, UK.

Hewitt J E and Norkko J (2007). Incorporating Temporal Variability of Stressors into Studies: An Example Using Suspension-Feeding Bivalves and Elevated Suspended Sediment Concentrations. *Journal of Experimental Marine Biology and Ecology*, **341**, 131–41.

Hewitt J E, Thrush S F, Gibbs M, Lohrer A and Norkko A (2006). Indirect Effects of *Atrina zelandica* on Water Column Nitrogen and Oxygen Fluxes: The Role of Benthic Macrofauna and Microphytes. *Journal of Experimental Marine Biology and Ecology*, **330**, 261–73.

Kline R (2011). *Principles and Practice of Structural Equation Modeling*, 3rd ed., Guilford Press, New York.

Kostylev V E, Todd B J, Fader G B J, Courtney R C, Cameron G D M and Pickrill R A (2001). Benthic Habitat Mapping on the Scotian Shelf Based on Multibeam Bathymetry, Surficial Geology and Sea Floor Photographs. *Marine Ecology Progress Series*, **219**, 121–37.

Landerretche O, Leiva B, Vivanco D and López I (2017). Welcoming Uncertainty: A Probabilistic Approach to Measure Sustainability. *Ecological Indicators*, **72**, 586–96.

Lejzerowicz F, Esling P, Pillet L, Wilding T A, Black K D and Pawlowski J (2015). High-Throughput Sequencing and Morphology Perform Equally Well for Benthic Monitoring of Marine Ecosystems. *Scientific Reports*, **5**, 13932.

Magorrian B H, Service M and Clarke W (1995). An Acoustic Bottom Classification Survey of Strangford Lough, Northern Ireland. *Journal of the Marine Biological Association of the United Kingdom*, **75**, 987–92.

Mathieu C, Lear G, Buckley T R, Hermans S M, Lee K C and Buckley H L (2020). A Systematic Review of Sources of Variability and Uncertainty in eDNA Data for Environmental Monitoring. *Frontiers in Ecology and Evolution*, **8**, 135.

Milner-Gulland E and Shea K (2017). Embracing Uncertainty in Applied Ecology. *The Journal of Applied Ecology*, **54** (6), 2063–8.

Pollman C D, Swain E, Bael D, Myrbo A, Monson P and Shore M (2017). The Evolution of Sulfide in Shallow Aquatic Ecosystem Sediments: An Analysis of the Roles of Sulfate, Organic Carbon, and Iron and Feedback Constraints Using Structural Equation Modeling. *Journal of Geophysical Research: Biogeosciences*, **122** (11), 2719–35.

Quinn G P and Keough M J (2002). *Experimental Design and Data Analysis for Biologists*, Cambridge University Press, Cambridge, UK.

Rodil I F, Attard K M, Norkko J, Glud R N and Norkko A (2019). Towards a Sampling Design for Characterizing Habitat-Specific Benthic Biodiversity Related to Oxygen Flux Dynamics Using Aquatic Eddy Covariance. *PloS One*, **14** (2), e0211673.

Schwinghamer P, Guigne J Y and Siu W C (1996). Quantifying the Impact of Trawling on Benthic Habitat Structure Using High Resolution Acoustics and Chaos Theory. *Canadian Journal of Fisheries and Aquatic Sciences*, **53**, 288–96.

Thomson J D, Weiblen G, Thomson B A, Alfaro S and Legendre P (1996). Untangling Multiple Factors in Spatial Distributions: Lilies, Gophers, and Rocks. *Ecology*, **77** (6), 1698–715.

Thrush S F, Halliday J, Hewitt J E and Lohrer A M (2008). The Effects of Habitat Loss, Fragmentation and Community Homogenization on Resilience. *Ecological Applications*, **18**, 12–21.

Thrush S F, Hewitt J, Cummings V J, Norkko A and Chiantore M (2010). β-Diversity and Species Accumulation in Antarctic Coastal Benthos: Influence of Habitat, Distance and Productivity on Ecological Connectivity. *PLoS One*, **5**, e11899.

Thrush S F, Hewitt J E and Lohrer A M (2012). Interaction Networks in Coastal Soft-Sediments Highlight the Potential for Change in Ecological Resilience. *Ecological Applications*, **22**, 1213–23.

Thrush S F, Hewitt J E, Norkko A, Nicholls P E, Funnell G A and Ellis J I (2003). Habitat Change in Estuaries: Predicting Broad-Scale Responses of Intertidal Macrofauna. *Marine Ecology Progress Series*, **263**, 113–25.

Uusitalo L, Tomczak M T, Müller-Karulis B, Putnis I, Trifonova N and Tucker A (2018). Hidden Variables in a Dynamic Bayesian Network Identify Ecosystem Level Change. *Ecological Informatics*, **45**, 9–15.

Wemheuer F, Taylor J A, Daniel R, Johnston E, Meinicke P, Thomas T and Wemheuer B (2020). Tax4Fun2: Prediction of Habitat-Specific Functional Profiles and Functional Redundancy Based on 16s rRNA Gene Sequences. *Environmental Microbiome*, **15** (1), 11.

Widdows J, Friend P L, Bale A J, Brinsley M D, Pope N D and Thompson C E L (2007). Inter-Comparison between Five Devices for Determining Erodability of Intertidal Sediments. *Continental Shelf Research*, **27** (8), 1174–89.

Wootton J T (1994). Predicting Direct and Undirect Effects: An Integrated Approach Using Experiments and Path Analysis. *Ecology*, **75**, 151–65.

Yee T W (2015). *Vector Generalized Linear and Additive Models: With an Implementation in R*, Springer-Verlag, New York.

Zajac R N, Lewis R S, Poppe L J, Twichell D C, Vozarik J and DiGiacomo-Cohen M L (2003). Responses of Infaunal Populations to Benthoscape Structure and the Potential Importance of Transition Zones. *Limnology and Oceanography*, **48**, 829–42.

Zar J H (1984). *Biostatistical Analysis*, Prentice-Hall, Upper Saddle River, NJ.

Zuur A F, Ieno E N and Elphick C S (2010). A Protocol for Data Exploration to Avoid Common Statistical Problems. *Methods in Ecology and Evolution*, **1**, 3–14.

PART III

Communities

CHAPTER 6

Describing assemblages and biodiversity of sediment-living organisms

6.1 Introduction

Benthic ecologists have seen the value in describing communities for over 100 years. These community descriptions were a way of identifying areas with similar groups of species and recognised both interactions between species and the response of species to changing environmental conditions as important in defining communities. This laid the foundation for mapping seafloor community type and documenting community change. Initially a community-characterising species was defined by Petersen (1924) as a species which was not seasonal and which exhibited numerical or biomass dominance. Overall, he defined seven species assemblages for the Kattegat area between Denmark and Sweden. These ideas were extended and seven major assemblage types defined, not only by their characterising species but also by the habitats (environments) in which they were found (Thorson, 1957). For example, the *Maldane-Ophiura sarsi* community was described as being found in soft muds in shallow estuaries and down to 300 m in the open sea, and characterised by *Maldane, Ophiura* and 14 other species. These days it is rare to see a study focussed on descriptions of assemblages by characterising species, especially to describe groups of species that may occur in specific habitats regardless of location. However, the concept of characterising assemblages by a subset of species is still used, mainly as a way to reduce the number of species that need to be discussed in studies. For example, risk assessment procedures based on the sensitivity of characterising species to stress (Tyler-Walters et al., 2009).

Part of the reason behind the swing away from assemblage descriptions was due to a debate about whether communities existed as discrete units (i.e. comprised of species whose ranges ended at the same position; Clements, 1916) or whether species occurred along gradients of environmental factors with different species having optima at different locations along the gradients (Gleason, 1926; Whittaker, 1975). Analysis of soft-sediment data showed the latter (Ugland & Gray, 1982), resulting in overlap between the groups of species observed at different sites.

The resultant practical definition of a community became a group of organisms, naturally occurring in a location and time, many of which interact in some way. A 'community' became a level of biological organisation and modern sampling designs and analytical procedures are used to show how community compositions change associated with stress, environmental conditions or the presence of specific species. Communities began to be described by multivariate analysis using ordination or clustering techniques that demonstrated how similar groups of species from different sites were, with a series of computer programmes developed to analyse them.

Ecology of Coastal Marine Sediments: Form, Function, and Change in the Anthropocene. Simon F. Thrush, Judi E. Hewitt, Conrad A. Pilditch and Alf Norkko, Oxford University Press (2021). © Simon F. Thrush, Judi E. Hewitt, Conrad Pilditch, and Alf Norkko.
DOI: 10.1093/oso/9780198804765.003.0006

At the same time, ecologists were developing other ways of describing the natural world and the species found in different locations. Numerous univariate methods to assess biodiversity were developed (that is, single value indices such as Margelef's richness, Pielou's evenness and the Shannon and Simpson indices) prior to 1975. Increasingly, ecologists working in soft sediments realised that they were not just describing natural communities but communities impacted by human activities, and differences between community ecology and biodiversity research became increasingly blurred. Worldwide biodiversity loss led to the need to better understand biodiversity including understanding effects of scale, more cost-effective means to assess biodiversity (including habitat and functional diversity), a better ability to predict biodiversity and its loss, and a greater need to understand the link between biodiversity and the ability of communities to recover from disturbance.

The rest of this chapter will focus on biodiversity in soft sediments, how it is studied, why it is important and the link between descriptions of biodiversity and communities.

6.2 What is biodiversity?

Biodiversity can be used to describe a number of different organisational scales from genes, genotypes, phenotypes, varieties, species and populations, all the way to communities, traits, habitats and ecosystems. But once we have decided on the organisational scale, we still have to decide what we mean by the term 'biodiversity'. As with many concepts in ecology, the meaning of biodiversity has not remained fixed over time.

There are many different univariate indices of diversity (Table 6.1), the simplest of which is species richness (or simply the number of species in a sample). As researchers tried to search for the one number that would characterise a community they invented multiple different indices that either represent different aspects of the community (level of dominance of the most abundant species or the number of rarer species or the degree of equitability in the distribution of individuals among species) or incorporated elements of both species richness and the distribution of individuals among species (Magurran, 2004). Bivariate definitions were also developed (Table 6.1) of which the most commonly

Table 6.1 Biodiversity can be defined in both simple and highly complex ways, related to community structure and function.

Definition	Single	Low complexity	Medium complexity	High complexity	High-complexity traits
Types	Richness	Richness/abundance univariate indices	Bivariate indices	Multivariate ordinations, classifications	Biological or functional traits
Examples	Number of species, Margelef's index	Shannon–Weiner, rarity, Simpson	Rank abundance, species abundance or species occurrence plots	Similarities and dissimilarities	Univariate measures (functional diversity, richness, number of traits), functional composition
Data used	Number of items	Number and abundance of items	Number of items and their abundance or frequency of occurrence	Number of items and their abundance or frequency of occurrence	Number of traits, abundance and frequency of occurrence within each trait
Strengths	Comparable between systems and regions	Comparable between systems and regions	Comparable between systems and regions	Track changes in quantity of individual items	Comparable between regions, track changes in quantity of individual traits
Weaknesses	Dependent on sample size, lack of information on abundance or frequency of items	Dependent on sample size, lack of information on individual items	Lack of information on individual items	Can only make comparisons between largely similar areas	Lack of information for categorising individual items

Adapted and extended from McGill et al. (2007). See section 6.3 for more information on individual measures.

used are the rank abundance, species occurrence distributions (SOD) and species abundance distributions (SAD; Figure 6.1). For all these univariate indices and bivariate measures, information on species identity is lost, thus comparisons can be made between systems that have no species in common (e.g. soft sediments, rocky reefs, forests) to help develop ecological generalities.

Increasingly, since the late 1970s, as computers became able to handle large numbers of species, ecologists have taken advantage of the ability of multivariate analyses to describe complex relationships (similarities and dissimilarities) within and between areas and communities and incorporated this within their concept and measurement of biodiversity. Multivariate methods retain both species

identities and abundances resulting in the ability to track changes in abundances of individual species. Comparisons can still be made between communities of widely different composition, as long as there is some overlap in species identities.

There are many different approaches that have been designed to suit different kinds of data and different questions (Legendre & Legendre, 1998). These methods are used: to demonstrate changes in community composition, associated with monitoring or manipulative experiments; to link changes in community composition with forcing factors; and to identify the level of (dis)similarity between one community and the next. These approaches are very powerful because they integrate across the responses of all species.

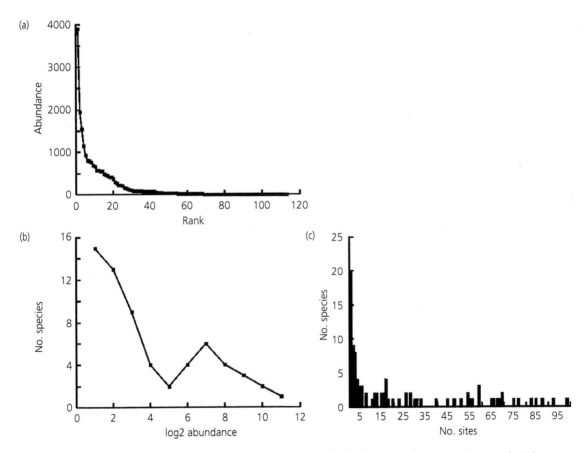

Figure 6.1 Rank abundance (A), species abundance (B) and species occupancy (C) plots demonstrate that most species occur either in low abundance or at few sites.

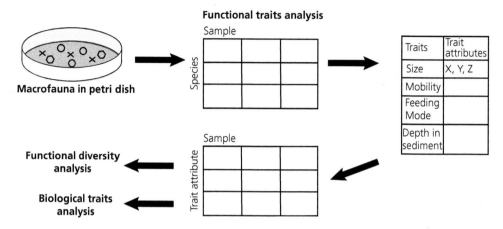

Figure 6.2 Species diversity can be amalgamated into functional diversity based on individual species' biological traits.

The need to be able to better explain and predict what the results of biodiversity loss may mean resulted in the development of the concept of 'functional' biodiversity and its use, particularly in soft sediments. Functional diversity uses information on the biological traits of species that affect how organisms respond to or modify their environment, interact with each other either within or between trophic levels, make choices about leaving suboptimum environments, recruit and disperse (Bremner et al., 2006; de Juan & Demestre, 2012; Törnroos & Bonsdorff, 2012; Villnäs et al., 2018) (Box 6.1). Although this topic has now been formalised as Biological Traits Analysis (BTA), it was preceded by many years of study on functional groups. Biological traits can either be left as individual traits or combined to create traits that represent specific ecosystem functions (Figure 6.2), for example the ability to bioturbate, increase oxygen concentrations at depth in the sediment or move organic material from the water column to the surface of the seafloor. A new set of biodiversity indices has been developed related to biological traits (Mouillot et al., 2013; see section 6.3).

Regardless of whether individual biological traits or combined functional traits are used, conversion of species to traits allows comparisons between locations where species pools differ (Hewitt et al., 2008). Descriptions of functional or trait diversity are particularly useful in soft-sediment systems as sampling usually incorporates multiple trophic levels and a mix of fauna and flora.

Box 6.1 Pros and cons of using traits

Trait analyses require caution for three major reasons. Firstly, results are dependent on the traits selected and therefore the traits used need to be selected to match the question of the study. For example, traits that make a species sensitive to a disturbance (response traits) are not necessarily those that will control recovery (recovery traits) and both of these are likely to differ from traits that are important for ecosystem functioning (effect traits). Secondly, we have limited knowledge of how many species behave (i.e. the traits they exhibit). Often studies have to use traits known at genera or family level but increasingly there are biological traits databases available, e.g. MarLIN (2006). Thirdly, many soft-sediment species exhibit a degree of plasticity in the way they behave, which may be environmentally driven or linked to species interactions. Whether plasticity occurs or not is most well known in feeding types, with many species varying their time spent between suspension/deposit feeding, scavenging/deposit feeding, etc. dependent on environmental variables such as currents or organic content. For example, *Macomona liliana* mainly deposit feeds but can raise its siphon into relatively still water and suspension feed. A possible way of dealing with both the latter issues is using fuzzy coding to represent the probability of a species exhibiting a certain type of behaviour. Thus, *Macomona* may be coded as predator (0), herbivore (0), suspension feeder (0.25) and deposit feeder (0.75). Similarly, a species that we knew little about other than that other species in that genera were either predators or scavengers may be coded as predator (0.5), scavenger (0.5). Obviously, if you know the environmental drivers controlling the switch, the species can be coded to one trait.

6.3 How do we measure biodiversity?

As mentioned previously, biodiversity can be represented at a number of different organisational scales. Measurements can focus on specific taxonomic groups (e.g. bacteria, molluscs, decapods) or particular sizes of organisms (meiofauna, macrofauna). Size-related measures are most common in soft sediments where studies will often focus on bacteria, microphytobenthos, meiofauna, macrobenthos or epibenthos. Studies that most commonly cross these size-related boundaries are macro-epibenthic studies or meio-macro studies. Descriptions based on individual phyla to orders are generally less common these days, although there is still often a division between kingdoms. For example, floral and faunal diversity are often not considered together (but see Thrush et al., 2011).

It is very rare to find a soft-sediment study of communities or biodiversity where all of the organisms can be described to species level. In part this is due to lack of taxonomic information on many family and genera, but also to practicalities and time/cost constraints. Rather than presenting information at a taxonomic level available across all observed taxa, most soft-sediment studies use variable levels, from species to phyla (e.g. nemerteans are often not identified lower than this). Thus richness is often defined as 'taxa richness' and taxa or operational taxonomic units are used to develop descriptions of composition. Many studies have looked at the effect of taxonomic resolution on results.

Multiple univariate indices have been developed over the years (for reviews see Hurlbert, 1971 and Magurran, 2004). Of these the most commonly used in soft sediments, and available in most software packages, remain number of taxa, Margelef's richness, Pielou's evenness and the Shannon–Weiner and Simpson indices. These can be used on data from any organisational scale including traits (e.g. functional trait richness and evenness; Thrush et al., 2017). Because any single trait can be expressed by a number of species (or taxa), these indices can also be calculated within traits (e.g. taxa richness of predators/scavengers; Hewitt et al., 2008). Other univariate indices have been developed specifically for trait diversity. Initially three indices were developed by Villeger et al. (2008): functional richness

(FRic), functional evenness (FEve) and functional divergence (FDiv). Later, Laliberté and Legendre (2010) extended these indices into a framework, implemented in the dbFD package in R (Laliberté et al., 2009), that includes four further indices: functional dispersion (FDis) (Laliberté & Legendre, 2010), Rao's quadratic entropy (Q) (Botta-Dukát, 2005), the community-level weighted means of trait values (CWM; Lavorel et al., 2008) and functional group richness (FGR) based on a posteriori functional classifications (Petchey & Gaston, 2006). Together these indices provide a strong representation of the ways that traits can be expressed in different communities and also the way they may respond to stressors (Mouillot et al., 2013). Another index that can be calculated on any organisational scale, for which relationships can be expressed on a Linnaean tree, is taxonomic distinctness, developed by Clarke and Warwick (1999) and widely used for a period of time in soft-sediment community analysis.

A final univariate measure of biodiversity relates to rarity. Rare species are those with low abundance and/or small range size (Gaston, 1994), are thus defined by either abundance or frequency of occurrence (see Glossary) and they may be rare in space or time (see Figure 6.1). Obviously, such definitions are strongly dependent on sampling effort and the spatial distribution of a species relative to the size of the sampling unit. Species that aggregate in patches the size or less than the size of the sampling unit are less likely to be observed as rare in abundance than those that are evenly dispersed (e.g. species whose feeding activities interfere with each other). Conversely, evenly dispersed species, generalists and widely dispersing species are less likely to be classified as rare in occurrence.

Multivariate measures rely mainly on the method used to define similarities and dissimilarities (association, similarity, distance and dependence coefficients) and the data transformation used on the data before these are applied. A large number of coefficients exist and their strengths and weaknesses for different types of data and questions are well summarised in Legendre and Legendre (1998). Generally soft-sediment ecologists tend to use four main distance measures: chi-square distances in correspondence analysis; Euclidean distances in principal component analysis; and Bray–Curtis or

Gower's dissimilarities in non-metric multidimensional scaling. The data transformations most commonly used are those that change the weight given to the most abundant taxa (e.g. square root, fourth root, log and presence absence). More complicated transformations allow down-weighting the role of rare species (ter Braak, 1987), vary the weighting given to rare and common taxa (Anderson et al., 2006) and allow ordination methods based on Euclidean distance measures to be applied despite problems generally associated with their use on ecological data (Legendre & Gallagher, 2001).

6.4 Biodiversity and scale

In the previous section we discussed some of the indices of diversity, but before we proceed further we need to consider the relationship between biodiversity, scale and sampling intensity. The issue of scale lies at the very centre of biodiversity assessment and, especially for soft sediments, the effect of scale is hard to separate from a discussion of sampling intensity and scale of sampling, as is highlighted in most other chapters. Biodiversity measures are generally categorised as occurring at a point

(alpha diversity), across a number of points or a region (gamma diversity) or variability/turnover between points or within a region (beta diversity). The scale at which alpha and gamma are calculated is chosen by the study or question, such that a single study may cover a number of different estimates of alpha and gamma at different scales (e.g. Figure 6.3).

Beta diversity is the most complicated of the measures and there are ongoing debates about how to measure it (Anderson et al., 2011; Legendre, 2014; Legendre & De Cáceres, 2013; Tuomisto & Ruokolainen, 2012)—and which indices to use. In soft-sediment systems, it is unusual to see beta diversity used with univariate indices such as Shannon–Weiner and Simpson or bivariate indices.

The alpha, beta and gamma aspects of biodiversity interact with sampling scale in a number of ways. One of the (if not the only) empirical laws in ecology (Lawton, 1999) is that species richness increases as a function of the area over which samples are collected (the Arrhenius power law; Arrhenius, 1921). At smaller scales the species accumulation curve (number of new species (or taxa) observed plotted against number of samples taken;

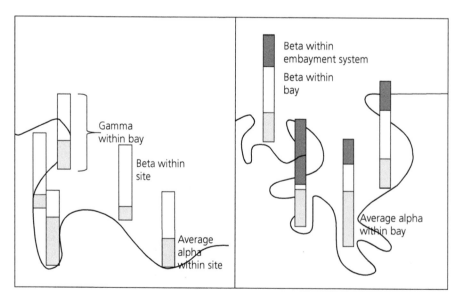

Figure 6.3 Alpha, beta and gamma diversity are scale-dependent definitions, with alpha diversity denoting the lowest scale. Here they are defined over a range of scales from a bay to an embayment to a regional scale.

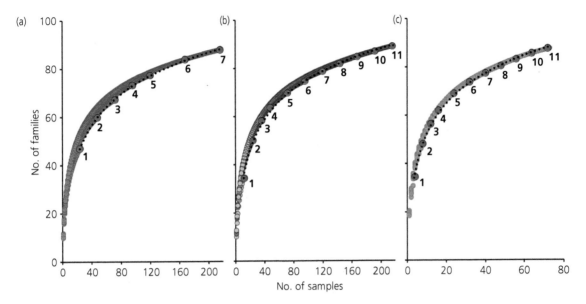

Figure 6.4 Regional taxon richness (at a family level) increases with increasing numbers of samples for the Mediterranean, accumulating across (a) subareas, (b) subareas and habitats and (c) subareas, habitats, and patchiness, from Bevilacqua et al. (2018). © 2017 Bevilacqua S, Ugland K I, Plicanti A, Scuderi D and Terlizzi A. *Ecology and Evolution* published by John Wiley & Sons Ltd Reprinted under Creative Commons Attribution 4.0 International (CC BY 4.0) licence.

Figure 6.4) is frequently used to assess whether sampling area and intensity are sufficient for estimating rarity, number of species, community composition and functional diversity. In this case, the assumption is that within a community/location the number of new species collected by each new sample will diminish until a true estimate can be obtained. Practically this is not always the case in marine sediments (Gray, 1981). There are a number of methods of predicting the curve (see Colwell, 2006 and Colwell & Coddington, 1994), available in Estimate S or Primer E, based on random selection of samples. Conversely, these curves can be used to determine the scales over which it is appropriate to calculate a representative measure of biodiversity (Nekola & White, 1999). For example, instead of developing the species accumulation curves based on a random selection process, the samples are selected either in terms of distance from each other (Thrush et al., 2010) or from sub-areas (Ugland et al., 2003). Distinct breaks in the curve where accumulation abruptly increases reveal areas/distances below which relative homogeneity in diversity occurs.

6.5 What have we learnt about biodiversity?

6.5.1 Rarity and commonness

An important commonality derived from studies of the way that abundances are distributed among species is that in most systems and organisational scales, most rank abundance plots (RA), species-abundance distributions (SADs) and species occupancy distributions (SODs) are left skewed (Figure 6.1). That is, most species occur in low abundances or at few sites; the majority of species in an area are rare. This holds true across systems from terrestrial to marine systems. In soft-sediment systems, similar patterns have been observed from continental shelves (e.g. in New Zealand 36% of taxa occurred at one site only (Ellingsen et al., 2007), in Norway 25% (Ellingsen, 2002), in Hong Kong 33% (Shin & Ellingsen, 2004), and in Fiji 42% (Schlacher et al., 1998)) and from deep sea (e.g. 36% in Eastern USA; Grassle & Maciolek, 1992).

The simple RA, SAD and SOD plots provide useful information on the degree to which one or a few species really dominate a community, how many rare species have been sampled or whether species are widely dispersed across sampling stations or only found at one or two specific places. These plots allow us to compare across very different communities and ask questions about diversity in different ecosystems. They plots have a long history in ecology, spawning many ideas on how individuals are distributed across species and how changes in the shape of the curve can indicate stress (Clarke, 1990; Gray & Mirza, 1979; Ugland & Gray, 1982).

Many studies have divided species into groupings around abundance and occurrence (e.g. 'core/common/widespread' versus 'rare/restricted range'), and used these classifications to infer general macroecological theory or community patterns (Coyle et al., 2013; Magurran & Henderson, 2003; Ugland & Gray, 1982; Ulrich & Ollik, 2004). Recently Gladstone-Gallagher et al. (2019) have suggested that widespread species are central to recovery dynamics. Rare species are expected to have a role in maintaining stability of functions when environmental conditions change (Walker et al., 1999; see Chapters 8 and 11). Another macroecological pattern, in this case related to commonness, is that spatial patterns in the richness of widespread species correlate well with spatial patterns of total species richness, total species turnover and environmental drivers (e.g. Jetz & Rahbek, 2002; Lennon et al., 2004; Morlon et al., 2008), although Hewitt et al. (2016) found no evidence for this in macrofaunal communities from intertidal areas of two harbours in New Zealand.

Finally, even the group of species that are neither common nor rare (the 'middle' group of fairly common species; Ugland & Gray, 1982) has been the subject of theory. Gray and Mirza (1979) observed that this group were the initial responders to organic enrichment of marine benthic communities. This observation was extended to postulate that 'over evolutionary time, species are moving up the geometric abundance scale as they become adapted to conditions or moving down when the fitness is reduced' (Ugland & Gray, 1982, p. 177).

6.5.2 General predictors of biodiversity

Other macroecological theories have focussed on large-scale patterns of biodiversity. Initially studies

looked across large gradients, e.g. latitude (matching terrestrial work), depth and temperature, and attempted to explain patterns with ecological theories around competition, predation, productivity and food resources. Latitudinal gradients similar to those found in terrestrial systems (decline in species richness towards arctic areas) have been reported for many marine groups (Levinton, 2001; Nybakken, 2001); however, specifically for soft-sediment systems, evidence is not conclusive. For example, latitudinal gradients have been reported for deep-sea benthos (Poore, 1993; Rex et al., 1993) but not for shallow-water nematodes (Boucher & Lambshead, 1995) or macrofauna (Kendall & Aschan, 1993). Shin and Ellingsen (2004, pp. 25–26) noted that 'few published surveys of tropical benthic communities have demonstrated very large numbers of species (Longhurst & Pauly, 1987)' and found lower numbers of species in their survey of Hong Kong waters compared to numbers observed off Norway (Ellingsen, 2001). Depth gradients have been documented from the classic work by Sanders in 1968, through to Carney (2005) and Stuart et al. (2017), although these are frequently attributed to variables that vary with depth, such as productivity.

Attention has also focussed on assessing diversity related to different habitats. For example, diversity is generally considered to be lower in soft muds than in coarse shelly sediments (Gray, 1974; Snelgrove, 1998). Current flow can also be important (Paterson & Lambshead, 1995) and in the deep sea, seeps can provide hot spots of diversity (Rogers et al., 2012). In soft-sediment systems, three-dimensional biogenic structures are important for providing habitats within which other species live and stabilising the seafloor (see Chapter 2). Although vegetative biogenic structures such as sea grasses are commonly accepted as supporting high biodiversity, sponge gardens, bryozoan reefs, rhodoliths, shellfish beds, tube worm habitats and oyster and mussel reefs are also important to biodiversity of soft sediments. Rhodoliths are found globally from the poles to the tropics in the photic zone extending from the intertidal zone to depths of over 200 m (Neill et al., 2015). NE Atlantic rhodolith beds have been found to harbour 30% of the total NE Atlantic algal flora, comparatively more than that found associated with kelp (c.10%) and seagrass (*c.*5%) (Peña et al., 2014). However, Hewitt et al. (2005) found small patches of broken shell to harbour diversity equivalent to many maerl or rhodolith beds. de Juan and Hewitt's (2011) study of intertidal flats of estuaries in New Zealand observed that it was generally the presence of habitat structure that was important and not specifically whether the structure was generated by seagrass, tube worms or shellfish. Conversely, species that bioturbate the sediment can destabilise the seafloor and disturb infauna (Norkko et al., 2006; Widdicombe et al., 2000), potentially decreasing diversity (Lohrer et al., 2008).

Generally, comparing biodiversity across studies is problematic due to different sampling unit sizes, sampling effort and taxonomic resolutions (Clarke, 1992; Gee & Warwick, 1996). Patterns may also depend considerably on what element of 'biodiversity' is measured (Clarke & Lidgard, 2000; Shin & Ellingsen, 2004; Rex & Etter, 2010). Within studies there is an acceptance that high variability in richness at a number of spatial scales occurs (Ellingsen & Gray, 2002; Shin & Ellingsen, 2004) related to environmental heterogeneity. In marine soft sediments, habitat heterogeneity is generally considered an important driver of biodiversity (Conner & McCoy, 1979; Ellingsen, 2002; Levin et al., 2001). In fact, Gray (1981) postulated that the decreased diversity at depths greater than 3000 m was a result of lack of environmental variation. Ellingsen and Gray (2002) demonstrated that habitat heterogeneity increased the slope of the relationship between area and species accumulation, such that predictions of biodiversity made without taking habitat heterogeneity into account were underestimates. Thrush et al. (2001) found macrofaunal diversity could be predicted by habitat heterogeneity in a nearshore embayment. At larger scales and at greater depths Compton et al. (2013) observed more diverse benthic assemblages in a region of high topographic and oceanographic complexity. Their study supports the hypothesis that continental margins are not monotonous mud slopes, as previously conceived (see review by Levin & Sibuet, 2012), but can be regionally diverse areas dependent on regional-scale environmental heterogeneity.

These findings do suggest a way forward for better predictions of biodiversity, design of biodiversity

studies and development of surrogates. Thrush et al. (2006) adapted the total species curve for increasing number of areas (Ugland et al., 2003) to deal with increasing numbers of habitats. Their results predicted that a greater variety of habitats, however defined, would hold higher species richness than a single habitat, even if the single habitat was considered to be a high species diversity habitat such as seagrass. However, their results also suggest that not only habitat heterogeneity is important but so is the proportion of habitat specialists to generalists within specific communities. Area and size of the habitat patch may also be important due to the interactions with hydrodynamics.

6.5.3 Connectivity, resilience and recovery

Biodiversity also plays a role in our understanding of connectivity and recovery from disturbances. As connectivity and recovery at small space and time scales underlie the degree to which broader-scale communities and ecosystems are predisposed to cross thresholds (Thrush et al., 2008), biodiversity may also play a role in defining the resilience of a location or community. This role is not fully understood, with some diversity–resilience theories predicting that more diverse communities should be more resilient and thus recover faster (Folke et al., 2004; Yachi & Loreau, 1999). However, leaving aside the concept of 'bad resilience' (communities degraded to the point of being comprised of opportunistic species are resilient to further degradation), a species-poor community should be able to reassemble faster than a species-rich community by chance alone.

Moreover, it has been argued that beta diversity can provide an estimate of ecological connectivity reflecting both hydrodynamic constraints and life history traits for all the species sampled at a particular site. Similarly, the local contribution to beta diversity (LCBD; Legendre & De Cáceres, 2013) represents the ecological uniqueness of a site.

However, dissimilarities among communities (beta diversity) can be considered to be a result of two different processes: species turnover (replacement) and differences in richness (species gain and loss) (Lennon et al., 2001; Patterson & Atmar, 1986). Recently methods have been proposed for decomposing beta diversity into these two components, and a third, nestedness. Generally, nestedness is expected to be a response to variation in habitats (e.g. complexity or habitat quality; Hylander et al., 2005), dispersal ability or habitat specialisation/tolerance (Heino et al., 2009). Baselga (2010) defines two aspects and methods for calculating both species turnover and nestedness (species loss). Podani and Schmera (2011) also provide methods for turnover and nestedness but include richness difference, i.e. the difference in the number of species found only at each site (see Carvalho et al., 2012 and Legendre, 2014 for a comparison of these methods). These different aspects of beta diversity are expected to drive recovery of sites after disturbances in different ways. For example, a site with high nestedness is a subset of a richer site and should recover fast as it is well connected to that richer site. However, this would obviously depend on how physically close the sites were and the time period over which dispersal occurs. In reality, the effect of differences in components of beta diversity on how soft-sediment communities respond to environmental change has not yet been widely studied.

Another source of theory/methods related to how community composition changes along environmental gradients is generated by the meta-community literature, in particular the pattern-based approach of 'elements of metacommunity structure' (EMS; Leibold & Mikkelson, 2002 and refined by Presley et al., 2010). This approach identifies various types of non-random species assemblages: continual change in species composition along environmental gradients without the formation of discrete assemblages (Gleason, 1926), nestedness (Patterson & Atmar, 1986) and mutually exclusive distributions (Diamond, 1975). Of course, soft-sediment communities rarely fit neatly into one type, depending on the degrees of nestedness, turnover and boundary clumping which tends to vary with environmental conditions, habitat type and taxonomic groups (Brustolin et al., 2019). The use of EMS in a broad-scale survey along part of the Finnish coast (Valanko et al., 2015) observed a pattern characteristic of meta-communities: the majority of species spanned a large portion of the environmental gradient and clumped at either end.

Josefson (2016) blended the EMS approach with a more usual soft-sediment analysis (variance partitioning) to understand drivers of communities in 16 estuaries around the Danish coast and observed a similar pattern to Valanko et al. (2015).

6.6 Close out

Understanding communities and biodiversity is a central topic in ecology and is in a constant state of change and development. Soft sediments with a diverse array of habitats and high species richness are ideal for testing and advancing theory. Sometimes new developments cycle back to re-examine aspects and theories of the past (such as the definitions of community by Gleason and Clements) and integrate them into new theory and understandings (meta-community theory). But we are far from learning all we need to understand about how soft-sediment communities develop and are modified, or about the processes that structure biodiversity and control its loss. Community and biodiversity research thus remains an exciting field for the future.

Much of the disagreement about the importance of particular patterns in biodiversity, let alone the factors that drive these patterns, is related to either the use of different biodiversity indices or inadequate sampling. An area of particular need for soft sediments is the development of effective measurement techniques and surrogates of biodiversity, and an understanding of what different indices and surrogates actually mean. Soft-sediment fauna and flora are hard to sample and even harder to observe in ways that allow understanding of their behavioural characteristics. Although new techniques may provide more extensive coverage and taxonomic resolution, without an understanding of how species behave and interact with each other and their environment our ability to understand community and biodiversity drivers will remain limited. In particular, surrogates for biodiversity need to be those that can be incorporated into understanding the role of biodiversity in ecosystem functioning (Chapter 9), for example, using functional traits.

Understanding how biodiversity scales both in space and in organisation is likely to continue to be a fruitful area of research. The finding that the most consistent predictor of biodiversity is large-scale habitat or environmental heterogeneity suggests that developing a hierarchical understanding of diversity may help predictions, including developing down-scaling laws and relationships for both space and organisational scales. However, given the multi-scale nature of most soft-sediment systems, consideration should also be given to investigating the potential to up-scale.

Importantly, it is imperative, when soft-sediment ecologists present studies of communities, biodiversity and drivers of either, that the scale of the study (Barton et al., 2013), the number of samples collected and the method by which biodiversity is expressed are explicitly stated. When making cross-study or cross-system comparisons, this information is key to being able to understand why differences may occur and developing generalities. When focussing on biodiversity alone it is important not to simply equate species richness with biodiversity, but to consider what a species does, how big it is and how many individuals there are.

References

Anderson M J, Crist T O, Chase J M, Velland M, Inouye B D, Freestone A L, Sanders N J, Cornell H, Comita L S, Davies K F, Harrison S, Kraft N J B, Stegen J C and Swenson N G (2011). Navigating the Multiple Meanings of β Diversity: A Roadmap for the Practicing Ecologist. *Ecology Letters*, **14**, 19–28.

Anderson M J, Ellingsen K E and McArdle B H (2006). Multivariate Dispersion as a Measure of Beta Diversity. *Ecology Letters*, **9**, 683–93.

Arrhenius O (1921). Species and Area. *Journal of Ecology*, **9**, 95–9.

Barton P S, Cunningham S A, Manning A D, Gibb H, Lindenmayer D B and Didham R K (2013). The Spatial Scaling of Beta Diversity. *Global Ecology and Biogeography*, **22** (6), 639–47.

Baselga A (2010). Partitioning the Turnover and Nestedness Components of Beta Diversity. *Global Ecology and Biogeography*, **19** (1), 134–43.

Bevilacqua S, Ugland K I, Plicanti A, Scuderi D and Terlizzi A (2018). An Approach Based on the Total-Species Accumulation Curve and Higher Taxon Richness to Estimate Realistic Upper Limits in Regional Species Richness. *Ecology and Evolution*, **8**, 405–15.

Botta-Dukát Z (2005). Rao's Quadratic Entropy as a Measure of Functional Diversity Based on Multiple Traits. *Journal of Vegetation Science*, **16** (5), 533–40.

Boucher G and Lambshead P J D (1995). Ecological Biodiversity of Marine Nematodes in Samples from Temperate, Tropical, and Deep-Sea Regions. *Conservation Biology*, **9** (6), 1594–604.

Bremner J, Rogers S I and Frid C L J (2006). Methods for Describing Ecological Functioning of Marine Benthic Assemblages Using Biological Traits Analysis (BTA). *Ecological Indicators*, **6**, 609–22.

Brustolin M C, Nagelkerken I, Moitinho Ferreira C, Urs Goldenberg S, Ullah H and Fonseca G (2019). Future Ocean Climate Homogenizes Communities across Habitats through Diversity Loss and Rise of Generalist Species. *Global Change Biology*, **25** (10), 3539–48.

Carney R S (2005). Zonation of Deep Biota on Continental Margins. In: Gibson R N, Atkinson R J A and Gordon J D M, eds. *Oceanography and Marine Biology: An Annual Review*, vol. 43, CRC Press, Boca Raton, FL, pp. 221–88.

Carvalho J C, Cardoso P and Gomes P (2012). Determining the Relative Roles of Species Replacement and Species Richness Differences in Generating Beta-Diversity Patterns. *Global Ecology and Biogeography*, **21** (7), 760–71.

Clarke A (1992). Is There a Latitudinal Diversity Cline in the Sea? *Trends in Ecology & Evolution*, **7** (9), 286–7.

Clarke A and Lidgard S (2000). Spatial Patterns of Diversity in the Sea: Bryozoan Species Richness in the North Atlantic. *Journal of Animal Ecology*, **69**, 799–814.

Clarke K R (1990). Comparisons of Dominance Curves. *Journal of Experimental Marine Biology and Ecology*, **138** (1), 143–57.

Clarke K R and Warwick R M (1999). The Taxonomic Distinctness Measure of Biodiversity: Weighting of Step Lengths between Hierarchical Levels. *Marine Ecology Progress Series*, **184**, 21–9.

Clements F E (1916). *Plant Succession: An Analysis of the Development of Vegetation*, Carnegie Institution of Washington, Washington, DC.

Colwell R K (2006). Estimates: Biodiversity Estimation, http://viceroy.eeb.uconn.edu/EstimateS.

Colwell R K and Coddington J A (1994). Estimating Terrestrial Biodiversity through Extrapolation. *Philosophical Transactions of the Royal Society of London*, **345**, 101–18.

Compton T J, Bowden D A, Pitcher C R, Hewitt J and Ellis N (2013). Biophysical Patterns in Benthic Assemblages across Contrasting Continental Margins off New Zealand. *Journal of Biogeography*, **40**, 75–8.

Conner E F and McCoy E D (1979). The Statistics and Biology of the Species–Area Relationship. *The American Naturalist*, **113**, 791–833.

Coyle J R, Hurlbert A H and White E P (2013). Opposing Mechanisms Drive Richness Patterns of Core and Transient Bird Species. *The American Naturalist*, **181** (4), e83–e90.

de Juan S and Demestre M (2012). A Trawl Disturbance Indicator to Quantify Large Scale Fishing Impact on Benthic Ecosystems. *Ecological Indicators*, **18**, 183–90.

de Juan S and Hewitt J (2011). Relative Importance of Local Biotic and Environmental Factors versus Regional Factors in Driving Macrobenthic Species Richness in Intertidal Areas. *Marine Ecology Progress Series*, **423**, 117–29.

Diamond J M (1975). The Island Dilemma: Lessons of Modern Biogeographic Studies for the Design of Nature Reserves. *Biological Conservation*, **7**, 129–46.

Ellingsen K E (2001). Biodiversity of a Continental Shelf Soft-Sediment Macrobenthos Community. *Marine Ecology Progress Series*, **218**, 1–15.

Ellingsen K E (2002). Continental Shelf Soft-Sediment Benthic Biodiversity in Relation to Environmental Variability. *Marine Ecology Progress Series*, **232**, 15–27.

Ellingsen K E and Gray J S (2002). Spatial Patterns of Benthic Diversity: Is There a Latitudinal Gradient along the Norwegian Continental Shelf? *Journal of Animal Ecology*, **71** (3), 373–89.

Ellingsen K E, Hewitt J E and Thrush S F (2007). Rare Species, Habitat Diversity and Functional Redundancy in Marine Benthos. *Journal of Sea Research*, **58**, 291–301.

Folke C, Carpenter S, Walker B, Scheffer M, Elmqvist T, Gunderson L and Holling C (2004). Regime Shifts, Resilience, and Biodiversity in Ecosystem Management. *Annual Review of Ecology, Evolution, and Systematics*, **35**, 557–81.

Gaston K J (1994). *Rarity*, Chapman and Hall, London, UK.

Gee J and Warwick R (1996). A Study of Global Biodiversity Patterns in the Marine Motile Fauna of Hard Substrata. *Journal of the Marine Biological Association of the United Kingdom*, **76** (01), 177–84.

Gladstone-Gallagher R V, Pilditch C A, Stephenson F and Thrush S F (2019). Linking Traits across Ecological Scales Determines Functional Resilience. *Trends in Ecology & Evolution*, **34**, 1080–91.

Gleason H A (1926). The Individualistic Concept of the Plant Association. *Bulletin of the Torrey Botanical Club*, **53**, 7–26.

Grassle J F and Maciolek N J (1992). Deep-Sea Species Richness: Regional and Local Diversity Estimates from Quantitiative Bottom Samples. *The American Naturalist*, **139** (2), 313–41.

Gray J S (1974). Animal-Sediment Relationships. *Oceanography and Marine Biology Annual Review*, **12**, 707–22.

Gray J S (1981). *The Ecology of Marine Sediments*, Cambridge University Press, Cambridge, UK.

Gray J S and Mirza F B (1979). A Possible Method for the Detection of Pollution-Induced Disturbance on Marine Benthic Communities. *Marine Pollution Bulletin*, **10**, 142–6.

Heino J, Mykrä H and Muotka T (2009). Temporal Variability of Nestedness and Idiosyncratic Species in Stream Insect Assemblages. *Diversity and Distributions*, **15** (2), 198–206.

Hewitt J E, Thrush S F and Dayton P D (2008). Habitat Variation, Species Diversity and Ecological Functioning in a Marine System. *Journal of Experimental Marine Biology and Ecology*, **366**, 116–22.

Hewitt J E, Thrush S F and Ellingsen K E (2016). The Role of Time and Species Identities in Spatial Patterns of Species Richness and Conservation. *Conservation Biology*, **30** (5), 1080–8.

Hewitt J E, Thrush S F, Halliday J and Duffy C (2005). The Importance of Small-Scale Habitat Structure for Maintaining Beta Diversity. *Ecology*, **86**, 1618–26.

Hurlbert S H (1971). The Nonconcept of Species Diversity: A Critique and Alternative Parameters. *Ecology*, **52** (4), 577–86.

Hylander K, Nilsson C, Gunnar Jonsson B and Göthner T (2005). Differences in Habitat Quality Explain Nestedness in a Land Snail Meta-Community. *Oikos*, **108** (2), 351–61.

Jetz W and Rahbek C (2002). Geographic Range Size and Determinants of Avian Species Richness. *Science*, **297**, 1548–51.

Josefson A B (2016). Species Sorting of Benthic Invertebrates in a Salinity Gradient—Importance of Dispersal Limitation. *PLoS One*, **11** (12), e0168908.

Kendall M A and Aschan M (1993). Latitudinal Gradients in the Structure of Macrobenthic Communities: A Comparison of Arctic, Temperate and Tropical Sites. *Journal of Experimental Marine Biology and Ecology*, **172** (1–2), 157–69.

Laliberté E and Legendre P (2010). A Distance-Based Framework for Measuring Functional Diversity from Multiple Traits. *Ecology*, **91**, 299–395.

Laliberté E, Legendre P and Shipley B (2009). Fd-Package Measuring Functional Diversity from Multiple Traits, and Other Tools for Functional Ecology. *R*.

Lavorel S, Grigulis K, McIntyre S, Williams N S, Garden D, Dorrough J, Berman S, Quétier F, Thébault A and Bonis A (2008). Assessing Functional Diversity in the Field – Methodology Matters! *Functional Ecology*, **22** (1), 134–47.

Lawton J H (1999). Are There General Laws in Ecology? *Oikos*, **84**, 177–92.

Legendre P (2014). Interpreting the Replacement and Richness Difference Components of Beta Diversity. *Global Ecology and Biogeography*, **23** (11), 1324–34.

Legendre P and De Cáceres M (2013). Beta Diversity as the Variance of Community Data: Dissimilarity Coefficients and Partitioning. *Ecology Letters*, **16** (8), 951–63.

Legendre P and Gallagher E D (2001). Ecologically Meaningful Transformations for Ordination of Species Data. *Oecologia*, **129**, 271–80.

Legendre P and Legendre L (1998). *Numerical Ecology*, Elsevier, Amsterdam.

Leibold M A and Mikkelson G M (2002). Coherence, Species Turnover, and Boundary Clumping: Elements of Meta-Community Structure. *Oikos*, **97** (2), 237–50.

Lennon J J, Koleff P, Greenwood J J D and Gaston K J (2001). The Geographical Structure of British Bird Distribtutions: Diversity, Spatial Turnover and Scale. *Journal of Animal Ecology*, **70**, 966–79.

Lennon J J, Koleff P, Greenwood J J D and Gaston K J (2004). Contribution of Rarity and Commonness to Patterns of Species Richness. *Ecology Letters*, **7** (2), 81–7.

Levin L A, Etter R J, Rex M A, Cooday A J, Smith C R, Pineda J, Stuart C T, Hessler R R and Pawson D (2001). Environmental Influences on Regional Deep-Sea Species Diversity. *Annual Review of Ecology and Systematics*, **32**, 51–93.

Levin L A and Sibuet M (2012). Understanding Continental Margin Biodiversity: A New Imperative. *Annual Review of Marine Science*, **4**, 79–112.

Levinton J (2001). *Marine Biology: Function, Biodiversity, Ecology*, Oxford University Press, Oxford, UK.

Lohrer A L, Chiaroni L D, Hewitt J E and Thrush S F (2008). Biogenic Disturbance Determines Invasion Success in a Subtidal Soft-Sediment System. *Ecology*, **89**, 1299–307.

Longhurst A R and Pauly D (1987). *Ecology of Tropical Oceans*, Academic Press, San Diego, CA.

McGill B J, Etienne R S, Gray J S, Alonso D, Anderson M J, Benecha H K, Dornelas M, Enquist B J, Green J L, He F, Hurlbert A H, Magurran A E, Marquet P A, Maurer B A, Ostling A, Soykan C U, Ugland K I and White E P (2007). Species Abundance Distributions: Moving beyond Single Prediction Theories to Integration within an Ecological Framework. *Ecology Letters*, **10**, 995–1015.

Magurran A E (2004). *Measuring Biological Diversity*, Blackwell Scientific, Oxford, UK.

Magurran A E and Henderson P A (2003). Explaining the Excess of Rare Species in Natural Species Abundance Distributions. *Nature*, **422**, 714–16.

MarLIN (2006). Biotic—Biological Traits Information Catalogue. *Marine Life Information Network*, Marine Biological Association of the United Kingdom, Plymouth [22 May 2020].

Morlon H, Chuyong G, Condit R, Hubbell S, Kenfack D, Thomas D, Valencia R and Green J L (2008). A General Framework for the Distance–Decay of Similarity in Ecological Communities. *Ecology Letters*, **11** (9), 904–17.

Mouillot D, Graham N A J, Villéger S, Mason N W H and Bellwood D R (2013). A Functional Approach Reveals Community Responses to Disturbances. *Trends in Ecology & Evolution*, **28** (3), 167–77.

Neill K, Nelson W, D'Archino R, Leduc D and Farr T (2015). Northern New Zealand Rhodoliths: Assessing Faunal and Floral Diversity in Physically Contrasting Beds. *Marine Biodiversity*, **45** (1), 63–75.

Nekola J C and White P S (1999). The Distance Decay of Similarity in Biogeography and Ecology. *Journal of Biogeography*, **26**, 867–78.

Norkko A, Rosenberg R, Thrush S F and Whitlatch R B (2006). Scale- and Intensity-Dependent Disturbance Determines the Magnitude of Opportunistic Response. *Journal of Experimental Marine Biology and Ecology*, **330**, 195–207.

Nybakken J W (2001). *Marine Biology: An Ecological Approach*, Benjamin Cummings, San Francisco, CA.

Paterson G and Lambshead P (1995). Bathymetric Patterns of Polychaete Diversity in the Rockall Trough, Northeast Atlantic. *Deep Sea Research Part I: Oceanographic Research Papers*, **42** (7), 1199–214.

Patterson B D and Atmar W (1986). Nested Subsets and the Structure of Insular Mammalian Faunas and Archipelagos. *Biological Journal of the Linnean Society*, **28** (1–2), 65–82.

Peña V, Barreiro R, Bárbara I, Cremades J, Díaz P, Pardo C, López L, Carro B, Piñeiro C and García V (2014). Maerl Beds in Galician Marine Protected Areas. How the Scientific Research Can Contribute to Their Management. *International Council for the Exploration of the Sea CM*, **3922 B**, 08.

Petchey O L and Gaston K J (2006). Functional Diversity: Back to Basics and Looking Forward. *Ecology Letters*, **9**, 741–58.

Petersen C G J (1924). A Brief Survey of the Animal Communities in Danish Waters. *American Journal of Science*, **7**, 343–54.

Podani J and Schmera D (2011). A New Conceptual and Methodological Framework for Exploring and Explaining Pattern in Presence – Absence Data. *Oikos*, **120**, 1625–38.

Poore G C (1993). Marine Species Richness. *Nature*, **361**, 597–8.

Presley S J, Higgins C L and Willig M R (2010). A Comprehensive Framework for the Evaluation of Metacommunity Structure. *Oikos*, **119** (6), 908–17.

Rex M A and Etter R J (2010). *Deep-Sea Biodiversity: Pattern and Scale*, Harvard University Press, Cambridge, MA.

Rex M A, Stuart C T, Hessler R R, Allen J A, Sanders H L and Wilson G D (1993). Global-Scale Latitudinal Patterns of Species Diversity in the Deep-Sea Benthos. *Nature*, **365** (6447), 636–9.

Rogers A D, Tyler P A, Connelly D P, Copley J T, James R, Larter R D, Linse K, Mills R A, Garabato A N and

Pancost R D (2012). The Discovery of New Deep-Sea Hydrothermal Vent Communities in the Southern Ocean and Implications for Biogeography. *PLoS Biology*, **10** (1), e1001234.

Sanders H L (1968). Marine Benthic Diversity: A Comparative Study. *The American Naturalist*, **102**, 243–82.

Schlacher T A, Newell P, Clavier J, Schlacher-Hoenlinger M A, Chevillon C and Britton J (1998). Soft-Sediment Benthic Community Structure in a Coral Reef Lagoon – the Prominence of Spatial Heterogeneity and 'Spot Endemism'. *Marine Ecology Progress Series*, **174**, 159–74.

Shin P K S and Ellingsen K E (2004). Spatial Patterns of Soft-Sediment Benthic Diversity in Subtropical Hong Kong Waters. *Marine Ecology Progress Series*, **276**, 25–35.

Snelgrove P V R (1998). The Biodiversity of Macrofaunal Organisms in Marine Sediments. *Biodiversity and Conservation*, **7** (9), 1123–32.

Stuart C T, Brault S, Rowe G T, Wei C L, Wagstaff M, McClain C R and Rex M A (2017). Nestedness and Species Replacement along Bathymetric Gradients in the Deep Sea Reflect Productivity: A Test with Polychaete Assemblages in the Oligotrophic North-West Gulf of Mexico. *Journal of Biogeography*, **44** (3), 548–55.

ter Braak C J F (1987). The Analysis of Vegetation-Environment Relationships by Canonical Correspondence Analysis. *Vegetatio*, **69**, 69–77.

Thorson G, ed. (1957). Bottom Communities (Sublittoral or Shallow Shelf). *Treatise on Marine Ecology and Paleoecology*, **1**, 461–534.

Thrush S F, Chiantore M, Asnagi V, Hewitt J, Fiorentino D and Cattaneo-Vietti R (2011). Habitat–Diversity Relationships in Rocky Shore Algal Turf Infaunal Communities. *Marine Ecology Progress Series*, **424**, 119–32.

Thrush S F, Gray J S, Hewitt J E and Ugland K I (2006). Predicting the Effects of Habitat Homogenization on Marine Biodiversity. *Ecological Applications*, **16**, 1636–42.

Thrush S F, Halliday J, Hewitt J E and Lohrer A M (2008). The Effects of Habitat Loss, Fragmentation and Community Homogenization on Resilience. *Ecological Applications*, **18**, 12–21.

Thrush S F, Hewitt J, Cummings V J, Norkko A and Chiantore M (2010). β-Diversity and Species Accumulation in Antarctic Coastal Benthos: Influence of Habitat, Distance and Productivity on Ecological Connectivity. *PLoS One*, **5**, e11899.

Thrush S F, Hewitt J E, Funnell G A, Cummings V J, Ellis J, Schultz D, Talley D and Norkko A (2001). Fishing Disturbance and Marine Biodiversity: Role of Habitat Structure in Simple Soft-Sediment Systems. *Marine Ecology Progress Series*, **221**, 255–64.

Thrush S F, Hewitt J E, Kraan C, Lohrer A M, Pilditch C A and Douglas E (2017). Changes in the Location of

Biodiversity–Ecosystem Function Hot Spots across the Seafloor Landscape with Increasing Sediment Nutrient Loading. *Proceedings of the Royal Society B*, **284**, 2016.2861.

Törnroos A M and Bonsdorff E (2012). Developing the Multitrait Concept for Functional Diversity: Lessons from a System Rich in Functions but Poor in Species. *Ecological Applications*, **22** (8), 2221–36.

Tuomisto H and Ruokolainen K (2012). Comment on "Disentangling the Drivers of β Diversity along Latitudinal and Elevational Gradients." *Science*, **335**, 1573.

Tyler-Walters H, Rogers S I, Marshall C E and Hiscock K (2009). A Method to Assess the Sensitivity of Sedimentary Communities to Fishing Activities. *Aquatic Conservation: Marine and Freshwater Ecosystems*, **19** (3), 285–300.

Ugland K I and Gray J S (1982). Lognormal Distributions and the Concept of Community Equilibrium. *Oikos*, **39**, 171–8.

Ugland K I, Gray J S and Ellingsen K E (2003). The Species-Accumulation Curve and Estimation of Species Richness. *Journal of Animal Ecology*, **72**, 888–97.

Ulrich W and Ollik M (2004). Frequent and Occasional Species and the Shape of Relative-Abundance Distributions. *Diversity and Distributions*, **10** (4), 263–9.

Valanko S, Heino J, Westerbom M, Viitasalo M and Norkko A (2015). Complex Metacommunity Structure for Benthic Invertebrates in a Low-Diversity Coastal System. *Ecology and Evolution*, **5** (22), 5203–15.

Villeger S, Mason N W H and Mouillot D (2008). New Multidimensional Functional Diversity Indices for a Multifaceted Framework in Functional Ecology. *Ecology*, **89**, 2290–301.

Villnäs A, Hewitt J, Snickars M, Westerbom M and Norkko A (2018). Template for Using Biological Trait Groupings When Exploring Large-Scale Variation in Seafloor Multifunctionality. *Ecological Applications*, **28**, 78–94.

Walker B, Kinzig A and Langridge J (1999). Plant Attribute Diversity, Resilience and Ecosystem Function: The Nature and Significance of Dominant and Minor Species. *Ecosystems*, **2**, 95–113.

Whittaker R H (1975). *Communities and Ecosystems*, Macmillan, New York.

Widdicombe S, Austen M C, Kendall M A, Warwick R M and Jones M B (2000). Bioturbation as a Mechanism for Setting and Maintaining Levels of Diversity in Subtidal Macrobenthic Communities. *Hydrobiologia*, **440** (1–3), 369–77.

Yachi S and Loreau M (1999). Biodiversity and Ecosystem Productivity in a Fluctuating Environment: The Insurance Hypothesis. *Proceedings of the National Academy of Sciences of the United States of America*, **96**, 1463–8.

CHAPTER 7

Biotic interactions

7.1 Introduction

Community processes have been a key area of research for ecologists. Generalisations on how species interact with each other and how these interactions are affected by the environment provide an understanding of how communities are structured, their functioning and dynamics and their likely response to changing environmental conditions and removal of species. Thus, information in this chapter underpins our ability to predict human impacts on soft-sediment systems (Chapter 11), changing biodiversity (Chapter 6) and the effects of climate change (Chapter 12). It is also necessary reading for those seeking to design sampling and studies that investigate a range of questions from predation to resilience through to marine protected area network design and conservation priorities. Similarly, an understanding of the role of natural disturbances and patch dynamics, mobility and the types of temporal variability that individual species can display is needed to place biotic interactions in context (see Chapter 3).

Although we think of a community as a level of biological organisation, it is hard not to give communities some geographic definition. In this chapter we will focus on the interactions between species (competition, predation, facilitation, parasitism and disease) that structure a community. We expand this to address the role of environmental and exogenous factors in moderating or amplifying the effects of biotic interactions and continue on to the description of interaction networks that seek to capture the emergent properties of a community. Finally, we also discuss connections between locations and communities that can affect how a community is assembled.

7.2 Bad neighbours, good neighbours and corporate raiders in soft sediments

7.2.1 Competition

Historically, two main categories of biotic interactions were assumed to control communities: competition and predation. The strong role of competition for space in structuring hard substrate communities inevitably led to the view that it would be similarly so in soft sediments. However, although it has been postulated that competition for space must occur as niches overlap, Gray (1981) suggested that this must be for short periods only as neither niche specialisation nor competitive exclusion are commonly observed phenomena in soft sediments (see also Baeta & Ramón, 2013; Peterson, 1979; Wilson, 1991). Instead, the theory of interference competition developed—one or more species preventing other species from exploiting a resource. Interference competition also underpinned the theory of trophic group amensalism (Rhoads & Young, 1970)—an interaction between sediment type, suspension feeders and deposit feeders. The theory goes that suspension-feeding larvae cannot establish in sediments with even moderate mud content, because deposit feeders disturb the surface and resuspend sediment particles. This was used to explain why suspension feeders are more prevalent than deposit feeders in sandy sediments and why the two feeding groups should not co-occur. Unfortunately, this simple model has not been supported by subsequent empirical tests; more complicated interactions are happening in marine sediments.

Very quickly another hypothesis arose, this time based on food limitation: exploitation competition.

Ecology of Coastal Marine Sediments: Form, Function, and Change in the Anthropocene. Simon F. Thrush, Judi E. Hewitt, Conrad A. Pilditch and Alf Norkko, Oxford University Press (2021). © Simon F. Thrush, Judi Hewitt, Conrad Pilditch, and Alf Norkko.
DOI: 10.1093/oso/9780198804765.003.0007

Levinton (1972) argued that suspension feeders must be opportunists as the quantity of plankton (which was expected to be their main food source) was temporally variable. Deposit feeders were expected to feed mainly on bacteria and this resource would be limited when sediment reworking resulted in large numbers of faecal pellets. Thus, suspension feeders would have overlapping niches and no specialisation, whereas deposit feeders would have evolved feeding specialisation. In fact the opposite is closer to the truth: many deposit feeders are bulk sediment feeders that just chew their way through the sediments (Lopez et al., 1989). Suspension feeders are much more specialised in terms of habitat use, the kinds of particles they can feed on and many are relatively large and slow-growing species—not a characteristic of opportunism.

This does not mean that competition for food does not occur. Resource partitioning between snails of different sizes based on size-dependent selection for ingested particles of different sizes was demonstrated by Fenchel and Kofoed (1976) and competition between suspension feeders can occur in dense beds (Svane & Ompi, 1993). In 1981, Levin (1981) demonstrated struggling over food and palp biting between adults of *Pseudopolydora paucibranchiata* (a tube-dwelling polychaete). A series of studies on blue crabs, *Callinectes sapidus*, in the Chesapeake Bay revealed that they were cannibalistic and in the presence of other adult crabs they would spend (well, actually waste) time putting on threatening displays and fighting, with a resultant decrease in feeding efficiencies (Clark et al., 1999). They were highly efficient at this, being able to detect other adults up to 4–5 m away! However, territoriality is not considered to be important for most soft-sediment species, despite multiple spatial studies conducted through the 1960s and 1970s that demonstrated even distributions. Most of the species which did exhibit such behaviour were tube or burrow dwellers and deposit feeders or predators (Levin, 1981).

7.2.2 Interference/inhibition

Some species (generally sediment mixers) produce environments that inhibit others. For example, soft-sediment urchins plough through the sediment and

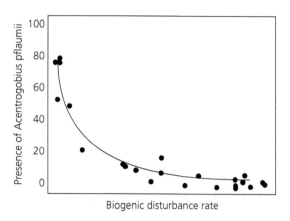

Figure 7.1   Presence of the burrowing goby, *Acentrogobius pflaumii*, is decreased with increasing biogenic disturbance rate (adapted from Figure 5, Lohrer et al., 2008).

in sufficient density can turn over sediment at 20,000 cm³ m⁻² d⁻¹. This has been documented to prevent the Asian goby from invading by collapsing burrow structures (Lohrer et al., (2008), Figure 7.1). Other species can destabilise the sediment surface, making it more prone to erosion (Chapter 2).

However, some cases of inhibition are very complicated. Adults of the bivalve *Macomona* live >2 cm below the sediment surface and feed on the surface by a siphon which moves around at the sediment–water interface. The frequent movement of the siphon can disturb small organisms trying to live their life in the top 2 cm of the sediment (see also examples in Chapter 3). But *Macomona* also have another behaviour that small organisms don't like (Woodin et al., 2016). Every now and then they cough! They clap their shells together and the resultant pressure wave can release porewater low in oxygen and high in ammonia across the sediment–water interface. Similar anoxic plumes have been recorded for the lugworm *Arenicola marina* (Volkenborn et al., 2010).

Often, however, inhibition might not be readily apparent, but still be expressed as legacy effects that affect the distribution of species. For example, large deposit feeders, such as lugworms (*Arenicola marina*), turn over the sediment and disrupt the burrows of small deposit-feeding amphipods (*Corophium volutator*). If the seafloor is disturbed and the large worms removed, the smaller and highly mobile amphipods will take advantage of the relaxing interference competition and increase in abundance, whereas the larger lugworms will take longer to recover (Norkko et al., 2006b). The moral is that there is a continuous response to a landscape of fear and competition.

7.2.3 Predation

Interestingly, none of the early theories emphasised the role of predation. Despite this, some of the earliest experiments in soft sediments (in the 1920s) were focussed on the benthos as fish food. However, the world of soft sediments is a multi-trophic-level world and exogenous predators (fish or birds or humans) are only one factor (Figure 7.2). Instead there are also three categories of predators that live within the soft-sediment system: epibenthic predators, infaunal predators and sublethal browsers. This usually creates multiple trophic levels within soft-sediment communities (Figure 7.3) often related to size, and two types of direct effects: firstly the consumption of prey and secondly the disturbance of the sediment as the predator excavates its food. Localised feeding pits, often formed by crustaceans, birds or fish, which expose other infauna to predation, can range from millimetres to metres in

length and cover large areas. This effect is not limited to natural disturbances; in fact, those created by human activities (e.g. dredging, sedimentation, hypoxia) not only expose infauna and epifauna to predation but often impact on the ability of the fauna to move away from predators.

Sublethal predation often consists of browsing on feeding tentacles, and this can have implications beyond the effect on the individual. For example, sublethal browsing that removes adult feeding tentacles of the spionid *Pseudopolydora kempi* decreases the effect of adults on juvenile conspecifics (Lindsay & Woodin, 1996). Siphon nipping of bivalves by shrimp or fish has been recorded as reducing the depth in the sediment at which the bivalves live, thus making them more available to bird predation (Cledón & Nuñez, 2010; De Goeij et al., 2001).

Numerous studies of predation have been undertaken, frequently using cages. Both inclusion and exclusion experiments have been conducted, with exclusion experiments being favoured on account of being more realistic and allowing for the effects on community structure to be revealed when the predator is at least temporarily removed. The numerous results of both kinds of experiment conducted in the 1970s–1990s were interesting, although not necessarily in the way the authors

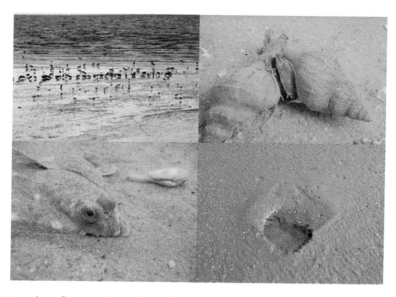

Figure 7.2 Predators on the seafloor range from endogenous (e.g. gastropods, polychaetes and crustaceans) to exogenous (birds and fish).

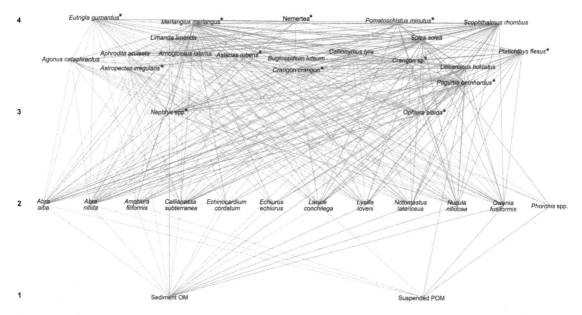

Figure 7.3 Soft-sediment systems are inherently comprised of multiple trophic levels. From Figure S2, Steger et al. (2019). Reproduced with permission granted by Inter research Science Publisher.

intended. In many studies, no direct negative effects of predators were identified (e.g. Raffaelli et al., 1989; Reise, 1977; Virnstein, 1979). Reviews of effects concluded that many predators are generalists and as a result their effect on community structure would be difficult to predict.

While this early work was considered to demonstrate a generally small role of predation in structuring soft-sediment communities, a review by Thrush (1999) highlighted a number of reasons why studies would underestimate effects: many predators are generalists (e.g. Mattila & Bonsdorff, 1988; Whitlatch, 1980); the presence of multiple trophic levels (Commito & Ambrose, 1985); the effect of sediment type on predation efficiency (Eggleston et al., 1992; Lipcius & Hines, 1986); and the ability of other species to provide refuges (see section 7.2.5).

However, probably the most important factor which relates specifically to soft sediments is the mobility of both predators and prey. This has two consequences: predators could be presented with such a high flux of prey that measurement of an effect is difficult (Gusmao et al., 2018); and studies that did not consider scale were likely to be ineffective (Thrush et al., 1999). Other aspects of scale that

have been found to be important were scales of habitat heterogeneity (Lipcius & Hines, 1986); dispersion of refuges (Keller et al., 2017; Lindberg et al., 1990); and effects of prey and predator aggregations on feeding rates (Schneider, 1992).

Building on these studies, Hines et al. (2009) conducted a test of how the opposing forces of aggregation and interference interact with the spatial distribution of prey patches to influence rates of prey consumption. They used detailed knowledge of blue crabs' ability to use chemosensory cues from clams to aggregate on prey patches up to 15 m apart, and their territorial aggression when closer than 5 m, and built a model to describe individual crab foraging efficiency as a function of distance between patches. The model predicted highest predation rates at lags between patches of 6.6 m, which was then validated by field experiments.

Micheli (1997) discovered complicated relationships between habitat use, density-dependent prey mortality and predators in a study of the bivalve, *Mercenaria mercenaria*, and the blue crab conducted over different habitat types at two times of the year. As well as finding differences in habitat usage and greater predation rates in high-density prey patches than in low-density patches, she found an important interaction between blue crab predation and the presence of large predatory birds. When these birds were abundant (in fall), crabs fed more in the salt marsh than in the sand flat habitat.

The important take-home message from these scale-related studies was that spatial and temporal heterogeneity were not noise that obscures processes but an active component of the predator–prey dynamic. These days many studies take advantage of heterogeneity to determine effects of predation (Baeta & Ramón, 2013; Seitz et al., 2017).

7.2.4 Parasitism and disease

Parasitism and disease are both biotic interactions, yet both are relatively infrequently studied. Sousa (1991, p. 821) declared: 'Despite the prevalence of parasites, their potential influence on soft-sediment communities is poorly understood.' At this time, the presence of parasites in numerous species had been documented but there was little work on effects on populations, with the exception of the wasting disease that nearly eliminated eelgrass populations from the North Atlantic during the 1930s. Seventeen years later, Kuris et al. (2008) claimed that parasites can sometimes account for a significant portion of the total biomass of natural ecosystems.

Microparasites that infect soft-sediment organisms include viruses, bacteria, fungi and a variety of protozoans. Macroparasites are most frequently digenetic trematodes but also include acanthocephalans, cestodes, nematodes and rhizocephalans. Microparasite infections are more likely to cause direct mortality than infections by macroparasites such as trematodes. Molluscs fully infected by larval trematodes have been documented as surviving up to several years (e.g. the salt marsh snail, *Cerithidea californica*; Sousa & Gleason, 1989). However, infections caused by either micro- or macro-parasites can result in high rates of mortality when associated with environmental stress (Glaspie et al., 2018; Magalhães et al., 2017; Sousa & Gleason, 1989).

Not surprisingly, the majority of parasite studies are on commercially important species: shellfish and crabs (e.g. Magalhães et al., 2017; Thrupp

et al., 2015). Frequently, studies on diseases focus on the prevalence and spread of the disease and not on the effects upon the fauna affected (Powell & Hofmann, 2015). Similarly, studies on parasites focus on identification, methods or diversity rather than their effects on their hosts (Blakeslee et al., 2012). For example, we know that cockles (*Cerastoderma edule* and *C. glaucum*) can display 59 different conditions being host to viruses, bacteria, fungi (including microsporidia), apicomplexa, amoeba, ciliophora, perkinsozoa, haplosporidia, cercozoa, turbellaria, digenea, cestoda, nematoda, crustacea and nemertea (Longshaw & Malham, 2013)!

The effect of parasites is frequently sublethal; they have been demonstrated to reduce fecundity, sometimes by castrating infections (Sindermann, 1970). There are also several recorded incidents of them altering burrowing behaviour and limiting growth (Longshaw & Malham, 2013; Ramilo et al., 2018). Parasites in cockles and *Macoma balthica* have been shown to increase time spent on the sediment surface, thus increasing the risk of predation by birds and epifaunal predators (Bonn, 1993; Swennen & Ching, 1974).

Obviously, a sublethal effect on reproduction can translate through from an individual to the population. This is particularly so if castration infections increase with host size/age (Peterson, 1983). Effects on individuals can translate to effects on populations in a number of ways. For example, a trematode infection causes a shift in habitat use by the Asian mud snail (*Batillaria cumingi*) to lower in the intertidal zone (Lefèvre et al., 2009). This has two further implications: increased access of the trematode to its secondary host; and movement of the infected snails away from the unaffected snails, reducing any competition for food. Similarly, the manipulative parasite *Microphallus papillorobustus* changed where the host, *Gammarus insensibilis*, lived (Ponton et al., 2005).

It is, however, important to realise that effects are not always detrimental. Magalhães et al. (2018) demonstrated an interaction between parasitism and arsenic that reduced the hosts' metabolism and thus decreased cellular damage from arsenic toxicity. Parasite biodiversity has even been suggested as an indicator of ecosystem health (Hudson et al., 2006).

Regardless of whether effects are detrimental to the host or not, changes to host behaviour, size or habitat use can easily translate through to changes to the ecosystem. If feeding, movements or spatial patterns are altered (Curtis, 1990), interspecies competitive interactions or predator–prey dynamics may also be altered. The ability to graze microphytobenthos may reduce, changing primary productivity within and on the seafloor. Most foodwebs do not include parasites; however, including these can increase the complexity considerably. Lafferty et al. (2008) demonstrate the ability of parasites to affect the chain length, connectance and robustness of foodwebs. In return food-web dynamics can also affect infectious disease dynamics.

7.2.5 The role of facilitation

Let us deal firstly with the obvious: large sedentary organisms protruding from the sediment surface. These protrusions, whether they be sponges, bivalves, rhodoliths, byrozoans or tube structures, alter the water flow over and around them, often creating areas of lower energy which other species can take advantage of to feed, rest or settle (see Chapter 2). This changing water flow also alters the flux of organics and sediment from the water column, providing different sediment characteristics and food resources for bacteria, microphytobenthos, meiofauna and larger species. In fact, if the species are suspension feeders they can considerably increase the flux of organics and sediment from the water column. The complex surface shape and three-dimensional structure also offer the opportunity for small species to gain refuge from larger predators whether they be fish, or epifaunal or infaunal invertebrates (Glaspie et al., 2018; Mendo et al., 2015; Talman et al., 2004). Depending on the species, they can extend the surface available for other species to live on (Figure 7.4).

However, species do not only increase vertical habitat by protruding from the sediment; there are numerous species that create structure by building burrows and creating mounds. Again, this three-dimensional structure increases the surface area available for colonisation and changes boundary flow conditions.

Although facilitation had been studied within the context of recovery from disturbance (see, e.g.

Figure 7.4 Habitat-structuring species: sponge, sea whips, hydroids and scallops on the seafloor in Terra Nova Bay, Antarctica (credit, Robotics lab, CNR, Genoa, Italy).

Gallagher et al., 1983 and Chapter 3), one of the first studies of facilitation within infaunal communities, according to Wayne Sousa, was by Thrush et al. (1992). Facilitation occurred with our friend *Macomona liliana*, but only at low adult densities, probably through the stimulation of microphyte production in non-eutrophic systems. At higher densities, *Macomona liliana* affects not only its own juveniles but also juveniles of other species, and indeed even small adults of other species, either by mixing the sediment or by promoting flow of anoxic water upwards through the sediment (see section 7.2.2). This illustrates how the nature of species interactions can vary dependent on density.

To this point, we have been talking about facilitation and inhibition as if they are fixed—a species either facilitates or inhibits or does not. However, that is not totally correct. For example, the effect of burrowers is strongly dependent on sediment type, with burrows remaining as distinct entities in muddy sediments much longer than in sandy sediment. In sandy sediment, burrowers have to continually dig new burrows or expend energy on maintenance (Needham et al., 2012). We will discuss this context-dependency more in the section below on exogenous modifiers.

7.2.6 Adult–juvenile interactions

Woodin (1976) really brought adult–larvae interactions to the forefront of community ecology in soft sediments. Her system involved interactions between suspension-feeding bivalves, tube builders and deposit feeders, with implications for these groups plus small burrowing polychaetes. This increased the number of groups of species that communities were built on and the addition of tube builders as sediment stabilisers recognised the fact that there are many species living in soft sediments that can produce a habitat that other species live within.

In the 1980s–1990s, the importance of post-settlement dispersal for soft-sediment systems began to be understood. That is, larvae may settle but then disperse repeatedly as juveniles. These adult–juvenile interactions may be quite different to adult–larval interactions. Both adult–larval and adult–juvenile interactions result in differences to spatial patterns in the abundance, population structure and biomass of both individual species and communities. For example, positive adult–juvenile interactions may result from adults providing a settlement surface, refuge from predation or changing the sediment type, but regardless result in areas containing a mix of life stages.

Negative adult–juvenile interactions result in juveniles only being able to colonise areas where adults are either not present or present in low abundances. Some species have evolved separate habitat requirements for adults and juveniles, possibly to deal with this. For others, juveniles colonise patches where adult densities are low. In both these cases the results can be large patches of single-age structures as mortality of adults needs to occur to allow the area to be available for juvenile colonisation (or larval settlement). Conversely, adult *Macomona liliana* are prey to stingrays which excavate pits sized 30 cm^{-1} m. Pits infill over 2 days to 1 week depending on the wave and current climate. However, the adults are relatively sedentary, leaving the infilled pits available to be colonised by juveniles, resulting in areas containing a mixture of life stages.

Environmental factors also can alter both the strength and direction of adult–juvenile interactions (Ahn et al., 1993; Dumbauld & Bosley, 2018; Olafsson, 1989). Sediment mud content may determine whether adults affect the porosity of the sediment and how easy it is for juveniles to burrow. Flow and waves are also likely to control the ability

of juveniles to settle into a favourable space or to leave an unfavourable space.

The importance of these adult–larval–juvenile interactions has profound implications for the distribution of species and the scale over which we need to think about population dynamics and community interactions. On the soft-sediment seafloor negative interactions can essentially create 'a floor of mouths' where settling juveniles can be smothered, physically disturbed or eaten. In these situations, the presence of shells or stones or other plants and animals on the seafloor can provide a refuge for new settlers, and thus play an important role in seafloor succession. Woodin's (1976) suggestion, that the collection of species living in a place arose mainly as a result of interactions between established individuals and newly settling larvae rather than between established individuals, linked neatly to theories around succession dynamics. If the soft-sediment world is comprised of patches at varying stages of recovery from disturbances as discussed in Chapter 3, succession dynamics are crucial to formation of communities.

7.3 Self-organisation

Studies of individual interactions revealed many varied in strength and so species could be connected directly and indirectly across communities. Many of the interactions we have discussed are negative, at least in terms of their immediate effects on competitors and prey. However, predation, competition and facilitation do not exist in isolation and through interaction networks many of these biotic interactions have much broader consequences on multiple species in the community, and often these interactions are more positive than negative.

Interactions between species, between different size classes of species and between different ecosystem components, particularly when the interactions occur at different scales, can result in apparently stable communities. Stable communities created by a combination of short-distance positive feedbacks and long-distance negative feedbacks are called self-organising. These have been demonstrated both for oyster reefs and for intertidal mudflat bathymetry (Rietkerk & Van de Koppel, 2008). For example, for the oyster reefs there are short-distance positive feedbacks between adults and juvenile oysters interacting with larger-spatial negative interactions between oysters and algae that maintain oyster banks interspersed with no-oyster areas.

However, self-organisation is a risky business (Box 7.1). Intertidal mudflats where regular spatial patterns develop as a result of an interaction between sedimentation and diatom growth can suddenly collapse when increased numbers of herbivores break the feedback loop (Weerman et al., 2012). Similarly, grazing on salt marshes by gastropods, in the presence of intense drought, results in die-off. The grazers accumulate along the edges of the die-off and amplify losses converting the marsh into mudflats (Silliman et al., 2005). Coco et al. (2006) showed how the relationship between hydrodynamics and density of the large pinnid bivalve *Atrina* could preserve densities in water that had high suspended sediment, but natural mortality of *Atrina* would break the benthic boundary flow conditions that protected the suspension feeder from the energetics required to feed on water with high suspended sediment loads.

7.4 Weak interactions

But not all seafloor communities are driven by an ecosystem engineer or a dominant 'key' species. Some communities are created and maintained by a series of relatively weak interactions between a number of species. For example, Angelini et al. (2018) demonstrate that multiple grazers regulate the productivity and drought resilience of south-eastern US salt marshes and that heterogeneity in physical stress and consumer density can dictate when and where top-down forcing is important.

Berlow (1999) demonstrated that 'weak' interactors can magnify spatial and temporal variation in community structure, such that effects on average are weak over broad scales but can still be strong in local contexts. Weak interactions are likely to involve a series of indirect effects. Generally, we know that indirect effects play an important role in structuring many marine communities; however, these effects are difficult to study, often being highly context-dependent. Wootton (2002) listed numerous challenges to understanding indirect effects which are still relevant today.

Whereas many studies focus on communities driven by ecosystem engineers and key species, we know much less about communities based around weak interactions and, in particular, whether they are more or less resilient to stressors. To some extent this may be because the area of ecosystem function has focussed on trophic and nutrient fluxes and biodiversity–ecosystem function experiments have focussed on effects of ecosystem engineers and key species. But it is also likely that it is largely a function of the difficulties in designing effective studies

---

Box 7.1 Exogenous modifiers

Many of the biotic interactions we observe can be altered by external variables. For example, post-settlement dispersal can be increased or interrupted by increased wave action, and burrowers or species feeding either on the sediment surface or in the water column can suspend feeding and burrowing activity during storms or intervals of high turbidity. Prolonged periods of high temperatures can dramatically alter microphytobenthos, macroalgal productivity and growth, biogeochemical cycles and movement of macrofauna and meiofauna. Importantly, facilitation is critically affected by stress, with events and large-scale fluctuations in environmental conditions able to increase or decrease facilitation, dependent on the sensitivity of the facilitating species to the stress. In some cases, facilitation may even be switched to inhibition (Balke et al., 2012; Norkko et al., 2006a).

Soft-sediment systems are also affected by exogenous biotic interactions, birds, fish and humans. Above the seafloor lies a complex web of predators and herbivores all supported at some stage by benthic species or processes. Sometimes the link between the benthos and these other species is one-way: a simple food source or complex 3D structure on the seafloor providing energy or predation refuges for small fish. However, interactions with these species may also be built into the biogeochemistry, population and community dynamics of the soft sediments. In 7.2.6 we discussed how stingray predation on a shellfish species actually helped maintain a population. In deep-sea systems, much of the organic carbon is delivered by settling zooplankton or phytoplankton, or by faecal pellets. Large megafauna such as whales provide organic carbon either as poo or as decaying carcasses settle on the seafloor. These can also create habitat structure (bones) and dig feeding pits.

and sampling in areas where engineers or key species are not clearly visible. Even the relationship between biodiversity and weak interaction communities remains generally unexplored, although Hammerschlag-Peyer et al. (2013) suggest that weak interactions are more common in species-rich systems and that they can buffer strong top-down effects. As 'complex networks of competing species may generate strong indirect interactions that can maintain diversity' (Aschehoug & Callaway, 2015, p. 452) it is important not to dismiss weak interactions. Thus, the view that Ecological Function Analysis 'needs to be performed for relatively few species or functions, making it a realistic way to improve conservation management' (Brodie et al., 2018, p. 840) is likely to create a biased result.

7.5 Connectivity

For soft sediments, a discussion of how communities are assembled and maintained is not complete without considering connectivity. What do we mean by connectivity? Any single soft-sediment location is connected in many ways.

There are connections that happen in place across time, flows within and between components such as sediment, bacteria, nutrients, oxygen, microphytes, meiofauna, macrofauna and macroflora, and these are discussed in many other specific chapters. There are also flows in place between organisational scales: individual to populations, species to communities, biotic habitats to communities and species and trophic links. Trophic links, expressed as foodwebs, have been frequently studied and theories posited around top-down (predator) and bottom-up (local resources) control (Seitz et al., 2017).

Although the quintessential marine trophic cascades occur in rocky reef systems, marine soft sediments also have the potential for cascades downwards to occur when an apex predator is removed (Estes et al., 2010). For example, removal of apex predatory sharks along the Atlantic seaboard coincided with increasing abundances of their mesopredatory prey species, particularly the cownose ray (*Rhinoptera bonasus*), and decreases in abundances of bivalves (*Mya arenaria, Crassostrea virginica, Mercenaria mercenaria, Argopecten irradians*

and *Macoma balthica*). Recently the potential for habitat or facilitation cascades has been raised, although given the numerous types of biotic interactions that make up a community, focussing on one aspect only will probably limit the understanding generated.

However, generally when we discuss connectivity we are talking about spatial connectivity. Even so, there are still various types of connectivity. Frequently, connectivity is simplified to hydrodynamic connections; can a particle (usually larvae) move in a parcel of water from one place to another? We have discussed in Chapter 3 how soft-sediment organisms exhibit different degrees of movement, often at different life stages. Importantly, most computer packages that model movement now allow for species behaviour: duration of larval stages, detecting suitable habitats, swimming up or down, etc.

7.6 Ecosystem interaction networks

We hope the discussion so far in this chapter demonstrates four things. Firstly, a soft-sediment system cannot be described based on foodwebs, feeding types or trophic levels alone, particularly when considering how it is likely to respond to change. Secondly, there are important links across space and time in how these systems operate that suggest that accumulating or averaging across these in models is unlikely to produce useful predictions. Thirdly, differences between adult and juvenile behaviour and habitat requirements mean network models must consider life stages. Fourthly, the lack of generalisations available for direction and intensities of interactions across feeding or other taxonomic groupings results in difficulties in producing valid groupings to be used in end-to-end ecosystem models such as Ersem, Ecopath and Atlantis.

However, ecosystem models are far from useless; ecosystem interaction networks are increasingly used to describe soft-sediment ecosystems (Figure 7.5), their resilience and the probability (and outcomes) of tipping points (Thrush et al., 2014). The benefit of these interaction networks is that we can piece together biotic interactions with their environmental drivers (from numerous studies) and evaluate

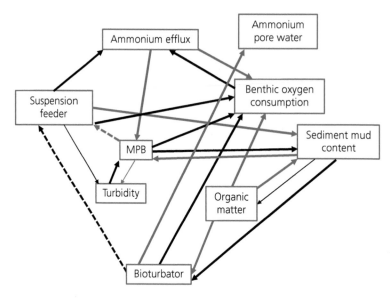

Figure 7.5 Increasingly ecosystem interaction networks are being developed to describe the function and structure of soft-sediment ecosystems. This one connects nutrient and oxygen dynamics with sediment characteristics, water clarity and benthic species traits of suspension feeding and bioturbating.

the relative importance of dynamic processes and changing biotic–abiotic interactions.

7.7 Close out

We have certainly learnt a lot over the years from studying the individual processes by which species interact. However, few studies move from investigating individual interactions to investigating how the community develops and is maintained—or changes. We also, as yet, know little about how the processes that affect the formation of assemblages control how stable the assemblage/community/system is and how resistant it is to endogenous change. But the question arises, do we really need to know the fine detail of these processes everywhere? Is it sufficient to understand the general processes? If we can recognise different types of species interactions and assemblages/systems, we may well be able to assess vulnerability to stressors and whether any change will be dramatic or gradual. Thrush et al. (2009) suggested that sensitivity of a key species to stress is likely to control resilience of the system. Many studies have demonstrated that positive feedbacks predispose

systems to tipping points if the positive feedback is sensitive to external drivers.

One point that must be recognised is the value of the large amount of natural history information that has been collected by studies on biotic interactions. Information on feeding behaviour (which is not always apparent from feeding apparatus structure), variations in mobility type and timing, living position, tendencies to aggregate or not; all this fed into the development of biological trait analysis (BTA). The uses of BTA in a number of fields range from enabling comparisons across species pool change, explaining the importance of biodiversity loss and the implications of impacts on ecosystem functioning (Oug et al., 2012), to assessing the sensitivity of habitats or areas to planned activities (Hewitt et al., 2018). However, there is a general acceptance in these uses that we do not yet have sufficient information behavioural characteristics of most species.

We mentioned that there has generally been little attention paid to the role of weak and indirect interactions. In the world of declining biodiversity and the recognition of non-random loss of species, research into these interactions is likely to become increasingly important. And in the world of increasing demand by the general public to restore what they see as degraded systems, ecologists will need to develop the understanding of how to build diverse, well-functioning communities and ecosystems.

References

Ahn I Y, Lopez G and Malouf R (1993). Effects of the Gem Clam *Gemma gemma* on the Early Post-Settlement Emigration, Growth and Survival of the Hard Clam *Mercenaria mercenaria*. *Marine Ecology Progress Series*, **99**, 61–70.

Angelini C, van Montfrans S G, Hensel M J, He Q and Silliman B R (2018). The Importance of an Underestimated Grazer under Climate Change: How Crab Density, Consumer Competition, and Physical Stress Affect Salt Marsh Resilience. *Oecologia*, **187** (1), 205–17.

Aschehoug E T and Callaway R M (2015). Diversity Increases Indirect Interactions, Attenuates the Intensity of Competition, and Promotes Coexistence. *The American Naturalist*, **186** (4), 452–9.

Baeta M and Ramón M (2013). Feeding Ecology of Three Species of *Astropecten* (Asteroidea) Coexisting on Shallow Sandy Bottoms of the Northwestern Mediterranean Sea. *Marine Biology*, **160** (11), 2781–95.

Balke T, Klaassen P C, Garbutt A, van der Wal D, Herman P M J and Bouma T J (2012). Conditional Outcome of Ecosystem Engineering: A Case Study on Tussocks of the Salt Marsh Pioneer *Spartina anglica*. *Geomorphology*, **153–154**, 232–8.

Berlow E L (1999). Strong Effects of Weak Interactions in Ecological Communities. *Nature*, **398**, 330–4.

Blakeslee A M H, Altman I, Miller A W, Byers J E, Hamer C E and Ruiz G M (2012). Parasites and Invasions: A Biogeographic Examination of Parasites and Hosts in Native and Introduced Ranges. *Journal of Biogeography*, **39** (3), 609–22.

Bonn A (1993). Epifaunal *Cerastoderma edule* L.-: Possible Effects of Trematode Parasites on the Burrowing Behaviour of Cockles. *PhD thesis, University of Wales, Bangor.*

Brodie J F, Redford K H and Doak D F (2018). Ecological Function Analysis: Incorporating Species Roles into Conservation. *Trends in Ecology & Evolution*, **33** (11), 840–50.

Clark M E, Wolcott T G, Wolcott D L and Hines A H (1999). Intraspecific Interference among Foraging Blue Crabs *Callinectes sapidus*: Interactive Effects of Predator Density and Prey Patch Distribution. *Marine Ecology Progress Series*, **178**, 69–78.

Cledón M and Nuñez J D (2010). Siphon Nipping Facilitates Lethal Predation in the Clam *Mesodesma*

mactroides (Reeve, 1854) (Mollusca: Bivalva). *Marine Biology*, **157** (4), 737–45.

Coco G, Thrush S F, Green M O and Hewitt J E (2006). The Role of Feedbacks between Bivalve (*Atrina zelandica*) Density, Flow and Suspended Sediment Concentration on Patch Stable States. *Ecology*, **87**, 2862–70.

Commito J A and Ambrose W G (1985). Multiple Trophic Levels in Soft-Bottom Communities. *Marine Ecology Progress Series*, **26**, 289–93.

Curtis L A (1990). Parasitism and the Movements of Intertidal Gastropod Individuals. *The Biological Bulletin*, **179** (1), 105–12.

De Goeij P, Luttikhuizen P C, Van der Meer J and Piersma T (2001). Facilitation on an Intertidal Mudflat: The Effect of Siphon Nipping by Flatfish on Burying Depth of the Bivalve *Macoma balthica*. *Oecologia*, **126** (4), 500–6.

Dumbauld B R and Bosley K M (2018). Recruitment Ecology of Burrowing Shrimps in US Pacific Coast Estuaries. *Estuaries and Coasts*, **41** (7), 1848–67.

Eggleston D B, Lipcius R N and Hines A H (1992). Density-Dependent Predation by Blue Crabs upon Infaunal Clam Species with Contrasting Distribution and Abundance Patterns. *Marine Ecology Progress Series*, **85**, 55–68.

Estes J A, Peterson C H and Steneck R S (2010). Some Effects of Apex Predators in Higher-Latitude Coastal Oceans. In: Terborgh J and Estes J A, eds. *Trophic Cascades: Predators, Prey, and the Changing Dynamics of Nature*, Island Press, Washington, DC, pp. 37–54.

Fenchel T and Kofoed L H (1976). Evidence for Exploitative Interspecific Competition in Mud Snails (Hydrobiidae). *Oikos*, **27**, 367–76.

Gallagher E D, Jumars P A and Trueblood D D (1983). Facilitation of Soft-Bottom Benthic Succession by Tube Builders. *Ecology*, **64**, 1200–16.

Glaspie C N, Seitz R D, Ogburn M B, Dungan C F and Hines A H (2018). Impacts of Habitat, Predators, Recruitment, and Disease on Soft-Shell Clams *Mya arenaria* and Stout Razor Clams *Tagelus plebeius* in Chesapeake Bay. *Marine Ecology Progress Series*, **603**, 117–33.

Gray J S (1981). *The Ecology of Marine Sediments*, Cambridge University Press, Cambridge, UK.

Gusmao J B, Lee M R, MacDonald I, Ory N C, Sellanes J, Watling L and Thiel M (2018). No Reef-Associated Gradient in the Infaunal Communities of Rapa Nui (Easter Island) – Are Oceanic Waves More Important than Reef Predators? *Estuarine, Coastal and Shelf Science*, **210**, 123–31.

Hammerschlag-Peyer C M, Allgeier J E and Layman C A (2013). Predator Effects on Faunal Community Composition in Shallow Seagrass Beds of the Bahamas. *Journal of Experimental Marine Biology and Ecology*, **446**, 282–90.

Hewitt J E, Lundquist C J and Ellis J (2018). Assessing Sensitivities of Marine Areas to Stressors Based on Biological Traits. *Conservation Biology*, **33**, 142–51.

Hines A H, Long W C, Terwin J R and Thrush S F (2009). Facilitation, Interference, and Scale: The Spatial Distribution of Prey Patches Affects Predation Rates in an Estuarine Benthic Community. *Marine Ecology Progress Series*, **385**, 127–35.

Hudson P J, Dobson A P and Lafferty K (2006). Is a Healthy Ecosytem One That Is Rich in Parasites? *Trends in Ecology & Evolution*, **21**, 381–5.

Keller D A, Gittman R K, Bouchillon R K and Fodrie F J (2017). Life Stage and Species Identity Affect Whether Habitat Subsidies Enhance or Simply Redistribute Consumer Biomass. *Journal of Animal Ecology*, **86** (6), 1394–403.

Kuris A M, Hechinger R F, Shaw J C, Whitney K L, Aguirre-Macedo L, Boch C A, Dobson A P, Dunham E J, Fredensborg B L, Huspeni T C, Lorda J, Mababa L, Mancini F T, Mora A B, Pickering M, Talhouk N L, Torchin M E and Lafferty K D (2008). Ecosystem Energetic Implications of Parasite and Free-Living Biomass in Three Estuaries. *Nature*, **454** (7203), 515–18.

Lafferty K D, Allesina S, Arim M, Briggs C J, De Leo G, Dobson A P, Dunne J A, Johnson P T J, Kuris A M, Marcogliese D J, Martinez N D, Memmott J, Marquet P A, McLaughlin J P, Mordecai E A, Pascual M, Poulin R and Thieltges D W (2008). Parasites in Food Webs: The Ultimate Missing Links. *Ecology Letters*, **11** (6), 533–46.

Lefèvre T, Lebarbenchon C, Gauthier-Clerc M, Missé D, Poulin R and Thomas F (2009). The Ecological Significance of Manipulative Parasites. *Trends in Ecology & Evolution*, **24** (1), 41–8.

Levin L A (1981). Dispersion: Feeding Behaviour and Competition in Two Spionid Polychaetes. *Journal of Marine Research*, **39**, 99–117.

Levinton J (1972). Stability and Trophic Structure in Deposit-Feeding and Suspension-Feeding Communities. *The American Naturalist*, **106**, 472–86.

Lindberg W J, Frazer T K and Stanton G R (1990). Population Effects of Refuge Dispersion for Adult Stone Crabs (Xanthidae, *Menippe*). *Marine Ecology Progress Series*, **66** (3), 239–49.

Lindsay S M and Woodin S A (1996). Quantifying Sediment Disturbance by Browsed Spionid Polychaetes: Implications for Competitive and Adult-Larval Interactions. *Journal of Experimental Marine Biology and Ecology*, **196**, 97–112.

Lipcius R N and Hines A H (1986). Variable Functional Responses for a Marine Predator in Dissimilar Homogeneous Microhabitats. *Ecology*, **67**, 1361–71.

Lohrer A L, Chiaroni L D, Hewitt J E and Thrush S F (2008). Biogenic Disturbance Determines Invasion Success in a Subtidal Soft-Sediment System. *Ecology*, **89**, 1299–307.

Longshaw M and Malham S K (2013). A Review of the Infectious Agents, Parasites, Pathogens and Commensals of European Cockles (*Cerastoderma edule* and *C. glaucum*). *Journal of the Marine Biological Association of the United Kingdom*, **93** (1), 227–47.

Lopez G, Taghon G and Levinton J (1989). *Ecology of Marine Deposit Feeders*, Springer-Verlag, New York.

Magalhães L, de Montaudouin X, Figueira E and Freitas R (2018). Interactive Effects of Contamination and Trematode Infection in Cockles Biochemical Performance. *Environmental Pollution*, **243** (B), 1469–78.

Magalhães L, de Montaudouin X, Freitas R, Daffe G, Figueira E and Gonzalez P (2017). Seasonal Variation of Transcriptomic and Biochemical Parameters of Cockles (*Cerastoderma edule*) Related to Their Infection by Trematode Parasites. *Journal of Invertebrate Pathology*, **148**, 73–80.

Mattila J and Bonsdorff E (1988). A Quantitative Estimation of Fish Predation on Shallow Soft-Bottom Benthos in SW Finland. *Kieler Meeresforschungen*, **6**, 111–25.

Mendo T, Lyle J, Moltschaniwskyj N and Semmens J (2015). Early Post-Settlement Mortality of the Scallop *Pecten fumatus* and the Role of Algal Mats as a Refuge from Predation. *ICES Journal of Marine Science*, **72** (8), 2322–31.

Micheli F (1997). Effects of Predator Foraging Behavior on Patterns of Prey Mortality in Marine Soft Bottoms. *Ecological Monographs*, **67** (2), 203–24.

Needham H R, Pilditch C A, Lohrer A M and Thrush S F (2012). Density and Habitat Dependent Effects of Crab Burrows on Sediment Erodibility. *Journal of Sea Research*, **76**, 94–104.

Norkko A, Hewitt J E, Thrush S F and Funnell G A (2006a). Conditional Outcomes of Facilitation by a Habitat-Modifying Subtidal Bivalve. *Ecology*, **87**, 226–34.

Norkko A, Rosenberg R, Thrush S F and Whitlatch R B (2006b). Scale- and Intensity-Dependent Disturbance Determines the Magnitude of Opportunistic Response. *Journal of Experimental Marine Biology and Ecology*, **330**, 195–207.

Olafsson E B (1989). Contrasting Influences of Suspension-Feeding and Deposit-Feeding Populations of *Macoma balthica* on Infaunal Recruitment. *Marine Ecology Progress Series*, **55**, 171–9.

Oug E, Fleddum A, Rygg B and Olsgard F (2012). Biological Traits Analyses in the Study of Pollution Gradients and Ecological Functioning of Marine Soft Bottom Species Assemblages in a Fjord Ecosystem. *Journal of Experimental Marine Biology and Ecology*, **432**, 94–105.

Peterson C H (1979). Predation, Competition, Exclusion and Diversity in the Soft-Sediment Benthic Communities of Estuaries and Lagoons. In: Livingston R J, ed. *Ecological Processes in Coastal and Marine Systems*, Plenum Press, New York, pp. 233–64.

Peterson C H (1983). A Concept of Quantitative Reproductive Senility: Application to the Hard Clam, *Mercenaria mercenaria* (L.)? *Oecologia*, **58** (2), 164–8.

Ponton F, Biron D G, Joly C, Helluy S, Duneau D and Thomas F (2005). Ecology of Parasitically Modified Populations: A Case Study from a Gammarid–Trematode System. *Marine Ecology Progress Series*, **299**, 205–15.

Powell E N and Hofmann E E (2015). Models of Marine Molluscan Diseases: Trends and Challenges. *Journal of Invertebrate Pathology*, **131**, 212–25.

Raffaelli D, Conacher A, McLachlan H and Emes C (1989). The Role of Epibenthic Crustacean Predators in an Estuarine Food Web. *Estuarine, Coastal and Shelf Science*, **28**, 149–60.

Ramilo A, Abollo E, Villalba A and Carballal M (2018). A *Minchinia mercenariae*-Like Parasite Infects Cockles *Cerastoderma edule* in Galicia (NW Spain). *Journal of Fish Diseases*, **41** (1), 41–8.

Reise K (1977). Predator Exclusion Experiments in an Intertidal Mud Flat. *Helgoländer wissenschaftliche Meeresuntersuchungen*, **30**, 263–71.

Rhoads D C and Young D K (1970). The Influence of Deposit-Feeding Organisms on Sediment Stability and Community Trophic Structure. *Journal of Marine Research*, **28**, 150–78.

Rietkerk M and Van de Koppel J (2008). Regular Pattern Formation in Real Ecosystems. *Trends in Ecology and Evolution*, **23**, 169–75.

Schneider D C (1992). Thinning and Clearing of Prey by Predators. *The American Naturalist*, **139**, 148–60.

Seitz R D, Lipcius R N and Hines A H (2017). Consumer versus Resource Control and the Importance of Habitat Heterogeneity for Estuarine Bivalves. *Oikos*, **126** (1), 121–35.

Silliman B R, van de Koppel J, Bertness M D, Stanton L E and Mendelssohn I A (2005). Drought, Snails and Large-Scale Die-Off of Southern U.S. Salt Marshes. *Science*, **310**, 1803–6.

Sindermann C J (1970). *Principal Diseases of Marine Fish and Shell Fish*, Academic Press, New York.

Sousa W P (1991). Can Models of Soft-Sediment Community Structure Be Complete without Parasites? *American Zoologist*, **31** (6), 821–30.

Sousa W P and Gleason M (1989). Does Parasitic Infection Compromise Host Survival under Extreme Environmental Conditions? The Case for *Cerithidea californica* (Gastropoda: Prosobranchia). *Oecologia*, **80** (4), 456–64.

Steger J, Pehlke H, Lebreton B, Brey T and Dannheim J (2019). Benthic Trophic Networks of the Southern North Sea: Contrasting Soft-Sediment Communities Share High Food Web Similarity. *Marine Ecology Progress Series*, **628**, 17–36.

Svane I and Ompi M (1993). Patch Dynamics in Beds of the Blue Mussel *Mytilus edulis* L.: Effects of Site, Patch Size, and Position within a Patch. *Ophelia*, **37**, 187–202.

Swennen C and Ching H L (1974). Observations on the Trematode Parvatrema Affinis, Causative Agent of Crawling Tracks of Macoma Balthica. *Netherlands Journal of Sea Research*, **8** (1), 108–15.

Talman S G, Norkko A, Thrush S F and Hewitt J E (2004). Habitat Structure and the Survival of Juvenile Scallops *Pecten novaezelandiae*: Comparing Predation in Habitats with Varying Complexity. *Marine Ecology Progress Series*, **269**, 197–207.

Thrupp T J, Pope E C, Whitten M M, Bull J C, Wootton E C, Edwards M, Vogan C L and Rowley A F (2015). Disease Profiles of Juvenile Edible Crabs (*Cancer pagurus* L.) Differ at Two Geographically-Close Intertidal Sites. *Journal of Invertebrate Pathology*, **128**, 1–5.

Thrush S F (1999). Complex Role of Predators in Structuring Soft-Sediment Macrobenthic Communities: Implications of Changes in Spatial Scale for Experimental Studies. *Australian Journal of Ecology*, **24** (4), 344–54.

Thrush S F, Hewitt J E, Dayton P K, Coco G, Lohrer A M, Norkko A, Norkko J and Chiantore M (2009). Forecasting the Limits of Resilience: Integrating Empirical Research with Theory. *Proceedings of the Royal Society B*, **276**, 3209–17.

Thrush S F, Hewitt J E, Parkes S, Lohrer A M, Pilditch C, Woodin S A, Wethey D S, Chiantore M, Asnaghi V, De Juan S, Kraan C, Rodil I, Savage C and Van Colen C (2014). Experimenting with Ecosystem Interaction Networks in Search of Threshold Potentials in Real-World Marine Ecosystems. *Ecology*, **95** (6), 1451–7.

Thrush S F, Lawrie S M, Hewitt J E and Cummings V J (1999). The Problem of Scale: Uncertainties and Implications for Soft-Bottom Marine Communities and the Assessment of Human Impacts. In: Gray J S, Ambrose W and Szaniawska A, eds. *Biogeochemical Cycling and Sediment Ecology*, Kluwer, Dordrecht, Netherlands, pp. 195–210.

Thrush S F, Pridmore R D, Hewitt J E and Cummings V J (1992). Adult Infauna as Facilitators of Colonization on Intertidal Sandflats. *Journal of Experimental Marine Biology and Ecology*, **159**, 253–65.

Virnstein R W (1979). Predation on Estuarine Infauna: Response Patterns of Component Species. *Estuaries*, **2**, 69–86.

Volkenborn N, Polerecky L, Wethey D S and Woodin S A (2010). Oscillatory Porewater Bioadvection in Marine Sediments Induced by Hydraulic Activities of *Arenicola marina*. *Limnology and Oceanography*, **55**, 1231–47.

Weerman E J, Van Belzen J, Rietkerk M, Temmerman S, Kefi S, Herman P M J and Van DeKoppel J (2012). Changes in Diatom Patch-Size Distribution and Degradation in a Spatially Self-Organized Intertidal Mudflat Ecosystem. *Ecology*, **93**, 608–18.

Whitlatch R B (1980). Patterns of Resource Utilization and Co-Existence in Marine Intertidal Deposit-Feeding Communities. *Journal of Marine Research*, **38**, 743–65.

Wilson W H (1991). Competition and Predation in Marine Soft-Sediment Communities. *Annual Review of Ecology and Systematics*, **21**, 221–41.

Woodin S A (1976). Adult-Larval Interactions in Dense Faunal Assemblages: Patterns of Abundance. *Journal of Marine Research*, **34**, 25–41.

Woodin S A, Volkenborn N, Pilditch C A, Lohrer A M, Wethey D S, Hewitt J E and Thrush S F (2016). Same Pattern, Different Mechanism: Locking onto the Role of Key Species in Seafloor Ecosystem Process. *Scientific Reports*, **6**, 26678.

Wootton J T (2002). Indirect Effects in Complex Ecosystems: Recent Progress and Future Challenges. *Journal of Sea Research*, **48** (2), 157–72.

CHAPTER 8

Temporal variations in benthic assemblages and processes

8.1 Introduction

Whereas human impacts are derived from multiple sources, the fundamental criterion for the health of our estuaries and coasts is the response of marine ecosystems. We need to understand their status and trends! Gathering and interpreting data on how soft-sediment ecosystems change over time is a fundamental issue that underpins both knowledge of ecological processes and monitoring to detect long-term changes in ecosystem health. Collecting data that allow us to separate variability associated with natural change from that resulting from human activities is particularly useful to avoid the unnecessary expense of over-defining trivial change which may be well below practical thresholds of concern.

Even if the questions or processes you are interested in are not outwardly temporal, it is important that you think about what can happen if patterns and relationships are not consistent. We live in a rapidly changing world and so we need to be much more conscious of ecosystem dynamics.

Inherently, we have a tendency to want what we observe to be unchanging. We accept daily, tidal and seasonal changes in primary productivity, environmental conditions, species behaviour and species abundances, while expecting the population, community or ecosystem to be stable—to exhibit temporal variations around an equilibrium. This would be so useful, allowing us to build up piecemeal studies of different aspects of a system over time about how the system works and to amalgamate data collected at different times into predictive models of, for example, species distribution. However, is this really how life is in soft sediments?

Here we discuss process studies and monitoring-the scales of temporal variability they have revealed and what they have taught us about soft sediments.

8.2 Studies of temporal variation

8.2.1 Process studies

One of the most important considerations when studying temporal variation is the length of time of the study and the frequency with which samples are taken. Unfortunately, there is no 'right' scale for studying particular processes. The scale of the study is generally chosen for practical purposes (within time, money or data collection method constraints) and the speed of the process of interest (see section 8.3.1).

However, the most common studies of temporal variability have been short-term studies of tidal or diurnal variation in behaviour, and sedimentary processes. These short-term studies have revealed important changes in sediment primary productivity and biogeochemistry (e.g. changes in nutrients, oxygen and organic and inorganic matter) and have been discussed in Chapters 1 and 2. Behavioural studies have revealed important variations in feeding, movement affecting bioturbation, and animals entering the water column. Corophid amphipods, for example, demonstrate diurnal dispersal, entering the water column during the night, either to return to the same location or to move to a different one. Intertidal cockles have endogenous feeding rhythms, feeding at high tide. These rhythms can persist when individuals are taken into constant laboratory conditions.

Ecology of Coastal Marine Sediments: Form, Function, and Change in the Anthropocene. Simon F. Thrush, Judi E. Hewitt, Conrad A. Pilditch and Alf Norkko, Oxford University Press (2021). © Simon F. Thrush, Judi Hewitt, Conrad Pilditch, and Alf Norkko. DOI: 10.1093/oso/9780198804765.003.0008

Studies on processes that occur over longer time scales have also been common, especially related to recovery from disturbance (Chapter 3), ecosystem functioning (Chapter 9), biodiversity (Chapter 6), dispersal and biotic interactions (Chapter 7), especially predator–prey dynamics. For example, seasonal studies of benthic–pelagic coupling document both seasonal and multiyear variation driven mainly by variability in the input of organic matter.

8.2.2 Monitoring

Data are not so commonly collected over time scales longer than five years, yet such data are very important in understanding system dynamics, long-term change, environmental impacts and assessing the status of ecological health or integrity. This is often driven by the lack of long-term funding and changes in the interests of individual researchers. Increasingly many countries fund long-term monitoring programmes to assess changes in the state of the marine environment (e.g. as is required in the People's Republic of China by their Environmental Protection Law; see Ministry of Ecology and Environment, 2018). This has resulted in a change in the types of long-term time series available, from time series driven by a single scientist to those run by institutions (Gray & Elliott, 2009). Such time series frequently include benthic macrofauna. Sometimes this is only a single important species, such as a commercially important shellfish, but more commonly a range of species or the full macrobenthic community are used to create measures of biodiversity or indices of health (Borja et al., 2000; Clark et al., 2019; Rodil et al., 2013).

Designing long-term monitoring programmes requires consideration of an extra set of questions (Lindenmayer et al., 2014; Hewitt & Thrush, 2019). Is the monitoring to answer a specific question (often related to a human activity in a specific location), or is it a more general State of the Environment monitoring (otherwise known as surveillance monitoring) programme? If the former, then use of environmental gradients and an emphasis on change along them over time are more robust (see Chapter 4). If surveillance monitoring is required, then there are another set of questions. Will a large area be sampled to create full spatial representation

of the area or will sentinel sites be chosen as likely to exhibit changes if anything affects the surrounding area? How frequently will sampling be conducted and how long will it be before changes can be detected with any reliability? Sampling multiple seasons within a year allows changes to be detected earlier and more robustly than when using annual (or longer) sampling. How often will analysis be undertaken on the data and adjustment of the monitoring be carried out? All these questions have cost–benefit implications and compromises generally occur between spatial (replication and spatial coverage) and temporal aspects (sampling frequency), as described in Chapter 5.

The change to institution/government-conducted monitoring programmes may (or may not) mean that they are more likely to be funded for long periods. However, such programmes frequently have problems related to consistency (e.g. methods changing, quality assurance and taxonomic resolution). Ellingsen et al. (2017) detail problems arising for the Norwegian monitoring and give a number of solutions for moving forwards.

Increasingly, techniques are being developed (or old techniques are finding a new use) to allow us to delve into the past. Records of aerial photographs and satellite imagery allow assessment of changes in nearshore habitats over recent times. Analysis of bivalve shells whether by photography or laser ablation can be used to give an indication of the environmental conditions under which the shell was formed. Combined with death assemblages and dating from sediment cores, this analysis can extend hundreds of years into the past.

8.3 What are the scales and types of temporal variability we observe?

The time scales at which temporal variability occurs in soft-sediment systems range from the very small scale of seconds, through seasonal and annual cycles to the longer time scales of centuries.

8.3.1 Fast and slow processes

The terminology of fast and slow processes was developed in the theoretical ecological literature and is central to concepts of resilience and regime shifts as we will discuss in section 8.5. But essentially, whether a process is fast or slow relates to the time scales over which variability occurs. Biogeochemical fluxes and processes dominated by bacteria are generally considered to be fast, with changes able to occur over seconds to minutes. Processes such as predation are variable at longer time scales, whereas processes such as facilitation may be variable over even longer time scales. However, size of the organism does not drive process speed. For example, the feeding of large suspension feeders occurs over short time scales. Importantly, most processes do not operate at a single time scale and so cannot easily be categorised as 'fast' or 'slow'. Primary productivity can fluctuate over minutes, but also fluctuate diurnally (with light), seasonally (with temperature) and interannually (with climate variability and ocean currents). And, of course, what is 'slow' and 'fast' can be considered relative to the scale of the study question—fluxes generated by bacteria may be slow if the study is interested in cellular reactions.

The importance to sampling and study design of understanding the time scales over which processes operate has been discussed in Chapter 4; here we will discuss some of the characteristics of organisms (generation times and behavioural characteristics)

that can affect the time scales over which processes vary.

Generally, the shorter the generation time (average time between generations), the smaller the time scale at which variability can occur. For macrofauna, although generation times may vary between months and many years, the interaction between space and time scales generated by feeding or mobility often defines the smallest temporal scale at which the species responds to the environment around it. Juvenile post-settlement mobility is a particularly strong characteristic in soft-sediment systems and is often accompanied by life-stage changes in habitat requirements, sensitivity to disturbances and contributions to ecosystem functioning, all of which result in observed temporal variability. When mobility changes with life stage, interactions between space and time result in differences in temporal resolution between life stages which can drive temporal patterns at longer time scales, e.g. seasonality and multi-year cycles (Figure 8.1).

Behavioural characteristics frequently of importance are: feeding and energy uptake rates; mobility; and reproductive frequency and type (e.g. brooding, eggs/seeds, vegetative, planktonic). Feeding and energy or nutrient uptake rates may determine growth rates and thus the smallest time period over

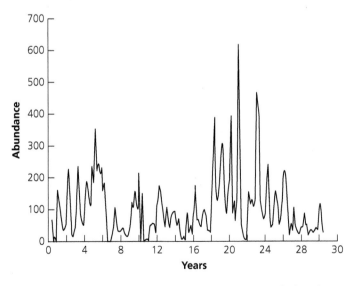

Figure 8.1 Abundance of a species frequently exhibits temporal variability at a variety of scales; in this figure from seasonal to 3 yearly to longer-term.

which standing stocks vary, particularly for percent cover of soft-sediment macroalgae or seagrasses. Organisms with continuous or semi-continuous reproduction with eggs, seeds, larvae or vegetative rafting often exhibit changes over shorter time periods than those with annual or single reproduction, which generally display seasonal or greater than annual cycles in abundance.

8.3.2 Seasonality

Longer studies of species abundances over periods of three-plus years emphasise seasonality driven by a variety of factors. Initially, seasonality was thought to be a function of seasonal differences in temperature and food resources. Thus, seasonality was expected to be higher in shallow areas compared to deep areas and in high latitudes compared to lower latitudes. In deep areas settlement and degradation processes were expected to smooth out pulses of organic detrital matter (Josefson, 1981) that could have driven seasonality of food resources; and in shallow areas large fluctuations in physio-chemical conditions driven by hydrodynamics were expected to produce strong variations in abundances of benthic species (Boesch & Rosenberg, 1981). These views changed over time, with seasonality in species abundances being observed in deep areas related to differences in settling phyto-detrital matter (Gooday, 2002). Similarly, although seasonal changes in species abundances are strongly marked in high latitudes, especially in areas where ice denudes the seafloor over winter, seasonality in abundances is also observed throughout temperate regions and the tropics. In the tropics a common driver of seasonality is monsoons, but in temperate regions the causes of seasonality generally relate to reproductive cycles which may occur at different times for different species throughout the year.

The few studies of processes that are replicated over time frequently demonstrate different results dependent on time. For example, a study of the effect of predation by birds and stingrays on a intertidal flat, conducted over a year, revealed seasonally variable and counterintuitive decreases in abundance of prey species driven by variation in predation rates, spatial scales of feeding, and predator species (shore birds in winter–spring (Cummings et al., 1997) and eagle rays in summer (Hines et al.,

1997)) and life stage-related changes in mobility of the prey. Processes can also be expected to vary in importance with spatial and temporal scale.

Not all species, however, demonstrate seasonality. Territoriality and slow growth combined with slow recruitment by semi-continuous or continuous reproduction may generate constant abundances (Gray & Christie, 1983). Hewitt and Thrush (2007) observed around 60% of common macrofaunal species did not exhibit seasonality of abundances in a large harbour in New Zealand.

8.3.3 Longer-term patterns and their drivers

Studies of 10-plus years are much less common in the soft-sediment literature, but usually reveal multiple interesting patterns. Sometimes these patterns are driven by variations in recruitment only; other patterns show increases in both recruitment and base-line abundances. Many bivalves demonstrate multi-year patterns in abundance driven by a year of very good recruitment followed by years of poor recruitment (Dame, 1996). Analysis by Gray and Christie (1983) of northern Europe long-term data sets demonstrated that many soft-sediment benthic species exhibited multi-year cycles in abundance of 5–7 (e.g. *Ciona intestinalis, Harmothoe sarsi, Echiurus echiurus, Pontoporeia* (now *Monoporeia*) *affinis* and *Paronychocamptus nana*) and 10–12 years (e.g. *Harmothoe, Echiurus* and *Monoporeia*). In an analysis of 20 years of data from a New Zealand harbour, Hewitt and Thrush (2007) observed up to 26% of the common taxa exhibited cycles in abundance of 3–9 years.

With longer time series of data we generally observe even longer cycles. Similar to mosaics of spatial variability, nested patterns begin to appear with smaller cycles superimposed over longer cycles, which are superimposed onto even longer cycles. Frequently, apparently unpredictable spikes and dips in abundances also appear, sometimes lasting for a brief period of time (less than two months), sometimes lasting for a year or more. Unpredictable spikes are most likely to occur with opportunistic species (short life-cycles and high fecundity) which can respond rapidly to changes in environmental conditions.

These longer time scales may be cycles occurring over 5–30 years as in the examples just given, or irreversible trends such as caused by invasive species or climate change. In fact, one important point to remember when analysing for trends is that any detected trends may in fact be an observation of a longer-term cyclic pattern. Another point is that the patterns observed will depend on the organisational resolution of the study. A species abundance at a specific location is likely to be more temporally variable than the temporal dynamics of the species within a harbour or the temporal dynamics of the community. Averaging over changes in species abundances can also occur when considering the temporal variability of the abundance of the functional trait to which the species belong.

Longer temporal patterns are likely to be the results of interactions between biotic processes occurring at different time scales (e.g. predator–prey dynamics), or biotic and environmental drivers (e.g. multi-year variations in recruitment, food resources or community composition). In the Baltic Sea, analysis of time series data of 40 years suggested that 5- to 7-year cycles in the abundance of *Monoporeia affinis* create similar-term cycles in the abundance of *Macoma baltica* as *Monoporeia* preys on *Macoma* larvae. Long-term cycles in bivalves have been attributed to relationships with temperature in both the Baltic and the Wadden Sea (Beukema et al., 2001). Rousi et al. (2013) attributed the long-term cycles in the abundance of *Monoporeia* to its sensitivity to increased temperatures. Gray and Christie (1983) suggested that known cycles in hydrographic conditions (salinity and temperature) may result in direct and indirect effects on abundances and reproductive rates of benthic organisms through changes in phytoplankton and zooplankton productivity, organic settlement rates and feeding rates of deposit feeders. In New Zealand, regression tree analysis of a 17-year data set observed lower densities of adults of the bivalve *Macomona liliana* with increasing temperatures in combination with lower rainfall in negative Southern Oscillation Index years (Hewitt et al., 2016a).

A major driver of long-term temporal patterns is global climate variability. The El Niño Southern Oscillation (ENSO) is probably the largest source of natural variability in the global climate. ENSO events occur irregularly but typically once every 3–6 years, nested within the 50+ year cycle of the Southern Oscillation (Allen et al., 1996). In the northern hemisphere the North Atlantic Oscillation (NAO) and the Atlantic Multidecadal Oscillation (AMO) are also important. The structuring power of these events for marine systems generally, and for soft sediments in particular, has been well demonstrated (Giani et al., 2012; Hagberg & Tunberg, 2000; Kröncke et al., 1998, 2019; Pacheco et al., 2012; Tunberg & Nelson, 1998). Many environmental variables are directly affected by these types of climate pattern events (e.g. oceanic circulation, water temperature, rainfall, and wind patterns and for the AMO the frequency and severity of Atlantic hurricanes; Zampieri et al., 2017). Cross-scale effects are therefore likely (Hewitt & Thrush, 2009b), especially in coastal areas where bathymetry and landscape features may produce small-scale spatial and temporal variability in wave energy and river discharges. Hewitt and Thrush (2009b) observed that a combination of the ENSO and smaller-scale location-specific environmental variables were able to explain up to 80% of the variability in species abundances over a 13-year period.

8.4 What we have learnt from temporal variability studies

Data collected over many years have been important in developing our understanding of how marine communities work.

8.4.1 What you see depends on when you are

Just as the relative importance of different processes varies over spatial scales, it also varies over temporal scales. More than that, the abundances of most species fluctuate over a range of spatial and temporal scales, driven by a combination of population dynamics and exogenous factors. Temporally fluctuating environmental conditions can affect the ability of many mobile species to disperse and persist within suitable habitats. Environmental fluctuations interacting with metapopulation dynamics can result in temporal variations in where and when species occur and in what abundances (McGeoch & Gaston, 2002).

Whereas the potential for spatial variability (e.g. differences in habitats, environmental gradients) is usually exploited to understand context dependency, the relative importance of processes and confounding or driving factors, this is rarely done for temporal variation (Box 8.1). Terrestrial studies of taxonomic groups frequently base biodiversity estimates on data accumulated over time; in marine systems the role of time is only just beginning to be explored (Kauppi et al., 2017; Mearns et al., 2015; Zajac et al., 2013).

Gray and Elliott (2009) went on to explore how changes in environmental conditions could alter the variability of community composition. Variability in community composition is most likely to occur driven by episodic events (e.g. storms and rainfall events in near-shore areas; Turner & Miller, 1991).

Such events may be irregular (e.g. hurricanes and earthquakes) or more regularly occurring (e.g. monsoon rains, ice scour). Episodic events can leave a legacy, resulting in changes in environmental conditions or dominant species that influence biodiversity and ecosystem function (Hernández-Miranda et al., 2014; Takami et al., 2017).

The connection between space and time also contributes strongly to the temporal inconsistencies observed in temporal dynamics across sites. These inconsistencies can result: from time lags in recruitment from source populations to sink areas (Figure 8.3); from historic differences in when a site

---

Box 8.1 Space and time are not easily separated

Seafloor spatial habitat mosaics are connected by water currents, energy flows and dispersal of organisms (Mahoney & Bishop, 2017). Although connectivity is most commonly considered to be a spatial process (Sheaves, 2009), the degree of connectivity is strongly temporally dependent, e.g. currents can be altered by climatic variability. Habitat utilisation is also temporally variable for many species. These complex interactions between space and time can result in different aspects of processes being important at different scales and in interactions between processes, space and time contributing to population and community dynamics (Figure 8.2).

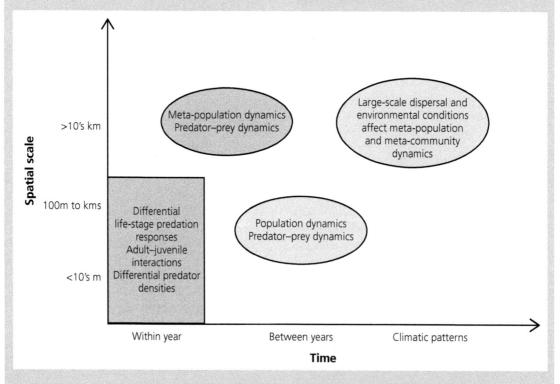

Figure 8.2 Processes vary in their importance across different space and time scales.

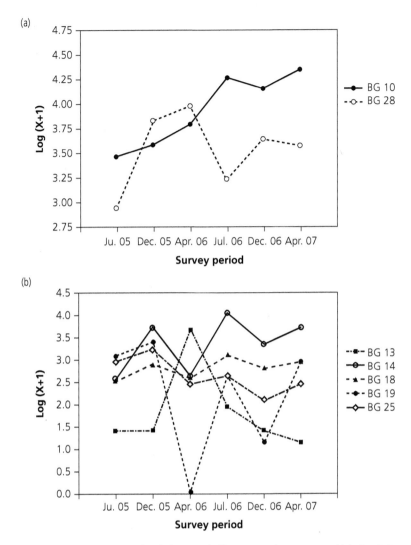

Figure 8.3 (a) Source–sink dynamics in the gastropod *Heleobia australis* (b) create asynchrony across multiple sites in Guanabara Bay, Rio de Janeiro, Brazil (from Figure 4, Echeverría et al., 2010). Reprinted under Creative Commons Attribution 4.0 International (CC BY 4.0) licence.

was initially colonised; from cross-scale interactions of environmental variables; or from a combination of environmental variability modified by species interactions (Dayton, 1989; Sellanesa et al., 2007) and density dependence (Fox & Morrin, 2001).

8.4.2 Variable species abundances, variable community composition?

As we observed earlier, different temporal patterns in a species' abundances can be observed across sites (asynchronicity), but this is also true for different species within a site. Recruitment rarely occurs for all species at the same time, and even when it does, not all species exhibit strong seasonality and some species exhibit juvenile or adult dispersal on a range of time scales. These factors add up to a community at a site being composed of a varying set of species with abundances that vary at temporally different scales. Hewitt et al. (2016b) noted that most species within their sites varied from being spatially rare to being common and vice versa (Figure 8.4).

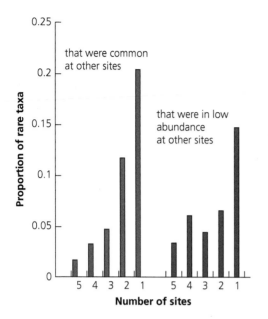

Figure 8.4 Macrofaunal species can be rare at one site in a large harbour but common at any number of other sites, even when the sites are located in similar habitat types.

These asynchronous changes in abundances may result in total abundances and species richness being less variable over time than for single species. However, for communities, Gray (1981) suggested that benthic macrofaunal communities display two types of behaviour: persistence, occurring when a key dominant controlled the range and numbers of other species; and oscillating patterns, which were more common. Recent work suggests that community composition may also be less variable over time when weak species interactions stabilise the community dynamics (Hewitt et al., 2010; Jacquet et al., 2016; van Nes & Scheffer, 2005) and, we would suggest, when communities form part of a meta-community. Conversely, when high densities of adults are required for successful recruitment (e.g. as may operate with broadcast spawners) or to maintain individual fitness (Allee effects), sudden disappearances of patches or even whole populations may occur.

The type of temporal variation observed is crucial to the insurance hypothesis. This hypothesis links maintaining biodiversity (in particular the number of rare species) to ensuring maintenance of ecological functions and resilience when environmental

conditions change (Magurran & Henderson, 2010; Naeem, 1998). The implication behind such insurance is that the rare species group contains species that either are more resistant to disturbance or environmental change than the current dominants or are better suited to respond to change in environmental/habitat characteristics. Studies have suggested insurance is more likely to occur where species' responses to environmental fluctuations are asynchronous, that is, not all species at a site have a similar sensitivity to a specific change in the environment (Loreau, 2010; Yachi & Loreau, 1999). A high degree of temporal structure in the identity of rare taxa within a meta-community structure increases the likelihood that rare species can provide insurance within the range of environmental conditions observed by the meta-community.

8.5 Resilience and tipping points

To this point we have dealt with natural variability; mainly reoccurring temporal patterns at varying scales. However, the coastal marine environment has been used by humans for many centuries and the number of uses and users are increasing with resultant changes in many aspects of these ecosystems. Sometimes the changes are slow and linear; other times they are not. While Chapter 11 will focus on human impacts and Chapter 12 on climate change, no discussion of temporal variability in coastal systems would be complete without including the essentially temporal concepts of 'resilience' and 'tipping points'.

Evidence is accumulating around the world that subtle but cumulative stressors can profoundly change the nature of marine ecosystems. These changes can variously be called tipping points, thresholds, state changes or regime shifts; they all represent a non-linear change in ecosystem state, often from a valued to a less valued one (Figure 8.5). Usually, but not always, they are accompanied by a change in how ecosystem components link together, and frequently they are unanticipated (surprises). Removal of the stress does not always result in recovery. These concepts are linked to 'ecological resilience'—the capacity of ecosystems to maintain their state in the face of stressors—and are likely to be related to decoupling of positive feedbacks

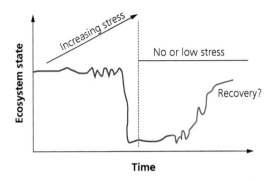

Figure 8.5 Idealised description of a tipping-point response to increasing stress and stress removal. The length of time and the path taken to recovery once stress was decreased were not the reverse of the degradation, demonstrating hysteresis has occurred.

between fast and slow processes (Rietkerk et al., 2004; van Nes et al., 2007). Evidence of abrupt changes in ecosystems, at time scales relevant to society, has been documented for coastal and shelf-depth systems (Casini et al., 2009; Edwards & Richardson, 2004; Ware & Thomson, 2005). Strangely, despite the number and intensity of human uses of estuarine and harbour systems (Altieri & Diaz, 2019; Lotze et al., 2006; McCormick-Ray, 2005; Thrush et al., 2013), there are few documentations of abrupt changes in these systems.

The suddenness, magnitude and apparent unexpectedness of these changes have led to considerable attention on how to determine that they are approaching (early warning signals). This has inevitably focussed on analysis of time series. Increased temporal variability of various components such as standard deviations and skewness has been identified in the general literature as precursors to sudden change (Carpenter & Brock, 2006; Guttal & Jayaprakash, 2009). In diverse systems, small-scale disturbances could lead to increasingly variable community composition. There is some support for this hypothesis from studies of variability in benthic communities along stress gradients (Warwick & Clarke, 1993). In a long-term study of the benthic communities off the NE coast of England, increased interannual variability and decreased multivariate stability were demonstrated (Warwick et al., 2002), but these did not occur until after abrupt changes in the water column associated with a regime shift. However, Hewitt and Thrush (2010) observed

increased interannual variability in single species abundances and community composition concurrent with changes in water quality.

Decreasing trends that would take the abundance of a key species or the rate of a key process to levels below those observed in natural cycles, or increasing deviance from natural levels, may indicate risk of regime shift (Thrush et al., 2009). The former could be determined by CUSUM, control charts and critical F-tests (e.g. Anderson & Thompson, 2004; Andersen et al., 2009) but the latter requires determining the 'natural' variability of a system, relative to variability in environmental forcing. This is by no means an easy task, requiring either long-term data prior to a change, or use of a 'reference' system. Finding a system that is similar enough to the system that has undergone change, yet has not been impacted by the same or another stressor(s), is generally very difficult. In a recent review of monitoring designs to detect tipping points, Hewitt and Thrush (2019) create guidelines that would allow the design of monitoring programmes capable of detecting tipping points.

Predicting approaching tipping points is complicated by the fact that changes to variance in the abundance of a species or a community parameter in response to stress are dependent on the spatial or temporal structure, e.g. interactions between spatial structure and disturbance regimes (Pascual & Guichard, 2005) or changes in aggregation of populations in relation to mean abundance (Murphy & Romanuk, 2012; Taylor, 1961). Over the last few years, ecologists have come to realise that it is not *increases* in variation that herald the tipping point, but any change in variation, be they increases or decreases (Hewitt & Thrush, 2009a; Kéfi et al., 2014).

Recovery after one of these abrupt changes may be affected by hysteresis (i.e. there may be a lag in recovery and the recovery trajectory followed may differ from the original pathway of change). Importantly, early work on stable states in soft-sediment systems observed that these systems comprise multiple stable states of differing degrees of resilience (Gray, 1977), suggesting that not only may the recovery trajectory be a different pathway but the system state 'recovered' may be different to the original.

8.6 Detecting patterns and trends

The combination of different processes and inter-actions producing different temporal patterns at different scales; the potential for resilience to create non-linearities; and the lack of knowledge about legacy and historic effects would at first glance seem to preclude understanding by ecologists. Certainly, this is the impression of the many who try to remove, or average over, temporal variability. However, temporal variability, similar to spatial variability, is rarely noise but instead contains a considerable amount of information.

Temporal variability can be directly studied and much information can be gathered on when temporal patterns occur and the scale at which they are manifested. This information often directly contributes to understanding the relative importance of different processes. For example, although monitoring can avoid recruitment times and thus appear to have a higher power to detect trends, stressors which affect recruitment will go undetected until the effect on recruitment has flowed through the population dynamics. Analytical techniques which allow both short and longer scales of variation to be directly incorporated in the analysis (e.g. ARIMA models) are useful tools in these cases, as is the use of co-variables such as Southern Oscillation values, precipitation and sea-surface temperatures (see Chapter 4 for a discussion of the use of co-variables).

It is particularly important to consider the temporal context of any study and match it with your question. There are numerous examples of inappropriate temporal contexts resulting in incorrect interpretations. For example, studying predation when juvenile fluxes of either prey or predators are high. Conversely, Kauppi et al. (2017) observed strong effects of the fauna on fluxes in spring when sediment organic content was high but very little effect in winter. Obviously modelling nutrient budgets based on values collected during spring could greatly miscalculate the annual nutrient fluxes. Whether unusual events have occurred before your study begins is obviously a concern and the state of the large climate drivers should also be taken into account (ENSO or NAO indices).

Temporal variability can be also incorporated into study designs. This is discussed mainly in Chapter 4, but should be particularly considered when: the time scales over which data have been collected differ between variables; variability is considered to be a response variable; and temporal variability is considered to be information rather than noise. When temporal variability is considered information, rather than testing for autocorrelation and removing it, techniques should be used to predict the wave length and amplitude of cycles (e.g. ARIMA, Fourier analysis, wavelets, and their multivariate versions: Principal Coordinates of Neighbourhood Matrix, Moran's eigenvector mapping).

In particular, we should refrain from expecting natural temporal variability to confound detection of anthropogenic effects or that human activities that impose stressors similar in size to natural variability will have little or no consequences. Human activities are superimposed on the wave of natural changes and frequently change the duration or frequency at which organisms are exposed to conditions near the limit of their ability to survive.

8.7 Close out

Although we are increasingly gaining long-term data, the present lack has resulted in a focus on spatial patterns and studies conducted at one time. Often this results in our assuming that the spatial patterns or relationships we observe are true over all time rather than being just a snapshot from a dynamic movie. This is particularly true for soft-sediment studies seeking to understand large-scale patterns because data are hard to collect. Although some evidence is accumulating that patterns not based on species identity may be generalisable over time, equally there is evidence that the importance of specific processes or species–environment relationships is time-dependent.

We maintain that soft sediments are temporally dynamic, exhibiting patterns at a variety of organisational, spatial and temporal scales, driven by numerous (interacting) processes. Sometimes repeatable cycles predominate, but sometimes they do not. Whereas it is obvious that what we observe about temporal patterns is highly dependent on our temporal scale of observation and the processes and species under study, it is not so intuitive that our observations are also dependent on the spatial scale of the study and the time it is conducted. However, an increasing number of studies replicated in time are allowing us to understand that temporal contexts exist—beyond the obvious related to recruitment and temperature. Importantly, the presence of seasonality and multi-year cycles observed for many processes, species abundances, biodiversity indices and community composition means that we need to be wary of any differences that we observe between samples collected at only a few times (Bonifácio et al., 2017; Compton et al., 2017; Schückel & Kröncke, 2013).

References

Allen R, Lindesay J and Parker D (1996). *El Nino Southern Oscillation and Climatic Variability*, CSIRO, Collingwood, Vic.

Altieri A H and Diaz R J (2019). Dead Zones: Oxygen Depletion in Coastal Ecosystems. In: Sheppard C, ed. *World Seas: An Environmental Evaluation*, 2nd ed., Academic Press, San Diego, CA, pp. 453–73.

Andersen T, Carstensen J, Hernandez-Garcia E and Duarte C M (2009). Ecological Thresholds and Regime Shifts: Approaches to Identification. *Trends in Ecology and Evolution*, 24, 49–57.

Anderson M J and Thompson A A (2004). Multivariate Control Charts for Ecological and Environmental Monitoring. *Ecological Applications*, 14, 1921–35.

Beukema J J, Dekker R, Essink K and Michaelis H (2001). Synchronized Reproductive Success of the Main Bivalve Species in the Wadden Sea: Causes and Consequences. *Marine Ecology Progress Series*, 211, 143–55.

Boesch D F and Rosenberg R (1981). Response to Stress in Marine Benthic Communities. In: Barrett G W and Rosenberg R, eds. *Stress Effects on Natural Ecosystems*, John Wiley & Sons, London, pp. 179–99.

Bonifácio P, Grémare A, Gauthier O, Romero-Ramirez A, Bichon S, Amouroux J-M and Labrune C (2017). Long-Term (1998 vs. 2010) Large-Scale Comparison of Soft-Bottom Benthic Macrofauna Composition in the Gulf of Lions, NW Mediterranean Sea. *Journal of Sea Research*, 131, 32–45.

Borja A, Franco J and Perez V (2000). A Marine Biotic Index to Establish the Ecological Quality of Soft-Bottom Benthos within European Estuarine and Coastal Environments. *Marine Pollution Bulletin*, 40, 1100–14.

Carpenter S R and Brock W A (2006). Rising Variance: A Leading Indicator of Ecological Transition. *Ecology Letters*, 9, 308–15.

Casini M J, Hjelm J, Molinero J-C, Lovgren J, Cardinale M, Bartolino V, Belgrano A and Kornilovs G (2009). Trophic Cascades Promote Threshold-Like Shifts in Pelagic Marine Ecosystems. *Proceedings of the National Academy of Sciences of the United States of America*, 106, 197–202.

Clark D E, Hewitt J E, Pilditch C A and Ellis J I (2019). The Development of a National Approach to Monitoring Estuarine Health Based on Multivariate Analysis. *Marine Pollution Bulletin*, 150, 110602.

Compton T J, Holthuijsen S, Mulder M, van Arkel M, Schaars L K, Koolhaas A, Dekinga A, ten Horn J, Luttikhuizen P C and van der Meer J (2017). Shifting Baselines in the Ems Dollard Estuary: A Comparison across Three Decades Reveals Changing Benthic Communities. *Journal of Sea Research*, 127, 119–32.

Cummings V J, Schneider D C and Wilkinson M R (1997). Multiscale Experimental Analysis of Aggregative Responses of Mobile Predators to Infaunal Prey. *Journal of Experimental Marine Biology and Ecology*, 216, 211–27.

Dame R F (1996). *Ecology of Bivalves: An Ecosystem Approach*, CRC Press, Boca Raton, FL.

Dayton P K (1989). Interdecadal Variation in an Antarctic Sponge and Its Predators from Oceanographic Climate Shifts. *Science*, 243, 151–60.

Echeverría C A, Neves R A F, Pessoa L A and Paiva P C (2010). Spatial and Temporal Distribution of the Gastropod *Heleobia australis* in an Eutrophic Estuarine System Suggests a Metapopulation Dynamics. *Natural Science*, 2, 860–7.

Edwards M and Richardson A J (2004). Impact of Climate Change on Marine Pelagic Phenology and Trophic Mismatch. *Nature*, 430, 881–4.

Ellingsen K E, Yoccoz N G, Tveraa T, Hewitt J E and Thrush S F (2017). Long-Term Environmental Monitoring for Assessment of Change: Measurement Inconsistencies over Time and Potential Solutions. *Environmental Monitoring and Assessment*, 189, 595.

Fox J W and Morrin P J (2001). Effects of Intra- and Interspecific Interactions on Species Responses to Environmental Change. *Journal of Animal Ecology*, 70, 80–90.

Giani M, Djakovac T, Degobbis D, Cozzi S, Solidoro C and Umani S F (2012). Recent Changes in the Marine Ecosystems of the Northern Adriatic Sea. *Estuarine, Coastal and Shelf Science*, 115, 1–13.

Gooday A J (2002). Biological Responses to Seasonally Varying Fluxes of Organic Matter to the Ocean Floor: A Review. *Journal of Oceanography*, 58 (2), 305–32.

Gray J S (1977). The Stability of Benthic Ecosystems. *Helgoländer Wissenschaftliche Meeresuntersuchungen*, 30, 427–44.

Gray J S (1981). *The Ecology of Marine Sediments*, Cambridge University Press, Cambridge, UK.

Gray J S and Christie H (1983). Predicting Long-Term Changes in Marine Benthic Communities. *Marine Ecology Progress Series*, **13**, 87–94.

Gray J S and Elliott M (2009). *Ecology of Marine Sediments: From Science to Management*, Oxford University Press, Oxford, UK.

Guttal V and Jayaprakash C (2009). Spatial Variance and Spatial Skewness: Leading Indicators of Regime Shifts in Spatial Ecological Systems. *Theoretical Ecology*, **2**, 3–12.

Hagberg J and Tunberg G B (2000). Studies on the Covariation between Physical Factors and the Long-Term Variation of the Marine Soft Bottom Macrofauna in Western Sweden. *Estuarine, Coastal and Shelf Science*, **50**, 373–85.

Hernández-Miranda E, Cisterna J, Díaz-Cabrera E, Veas R and Quiñones R A (2014). Epibenthic Macrofaunal Community Response after a Mega-Earthquake and Tsunami in a Shallow Bay off Central-South Chile. *Marine Biology*, **161** (3), 681–96.

Hewitt J, Thrush S, Lohrer A and Townsend M (2010). A Latent Threat to Biodiversity: Consequences of Small-Scale Heterogeneity Loss. *Biodiversity and Conservation*, **19**, 1315–23.

Hewitt J E, Ellis J I and Thrush S F (2016a). Multiple Stressors, Nonlinear Effects and the Implications of Climate Change Impacts on Marine Coastal Ecosystems. *Global Change Biology*, **22**, 2665–75.

Hewitt J E and Thrush S F (2007). Effective Long-Term Monitoring Using Spatially and Temporally Nested Sampling. *Environmental Monitoring and Assessment*, **133**, 295–307.

Hewitt J E and Thrush S F (2009a). Do Species' Abundances Become More Spatially Variable with Stress? *Open Ecology Journal*, **2**, 37–46.

Hewitt J E and Thrush S F (2009b). Reconciling the Influence of Global Climate Phenomena on Macrofaunal Temporal Dynamics at a Variety of Spatial Scales. *Global Change Biology*, **15**, 1911–29.

Hewitt J E and Thrush S F (2010). Empirical Evidence of an Approaching Alternate State Produced by Intrinsic Community Dynamics, Climatic Variability and Management Actions. *Marine Ecology Progress Series*, **413** (Special issue on Threshold Dynamics in Marine Benthic Ecosystems), 267–76.

Hewitt J E and Thrush S F (2019). Monitoring for Tipping Points in the Marine Environment. *Journal of Environmental Management*, **234**, 131–7.

Hewitt J E, Thrush S F and Ellingsen K E (2016b). The Role of Time and Species Identities in Spatial Patterns of Species Richness and Conservation. *Conservation Biology*, **30** (5), 1080–8.

Hines A H, Whitlatch R B, Thrush S F, Hewitt J E, Cummings V J, Dayton P K and Legendre P (1997). Nonlinear Foraging Response of a Large Marine Predator to Benthic Prey: Eagle Ray Pits and Bivalves in a New Zealand Sandflat. *Journal of Experimental Marine Biology and Ecology*, **216**, 211–28.

Jacquet C, Moritz C, Morissette L, Legagneux P, Massol F, Archambault P and Gravel D (2016). No Complexity–Stability Relationship in Empirical Ecosystems. *Nature Communications*, **7**, 12573.

Josefson A B (1981). Persistence and Structure of Two Deep Macrobenthic Communities in the Skagerrak (West Coast of Sweden). *Journal of Experimental Marine Biology and Ecology*, **50**, 63–97.

Kauppi L, Norkko J, Ikonen J and Norkko A (2017). Seasonal Variability in Ecosystem Functions: Quantifying the Contribution of Invasive Species to Nutrient Cycling in Coastal Ecosystems. *Marine Ecology Progress Series*, **572**, 193–207.

Kéfi S, Guttal V, Brock W A, Carpenter S R, Ellison A M, Livina V N, Seekell D A, Scheffer M, van Nes E H and Dakos V (2014). Early Warning Signals of Ecological Transitions: Methods for Spatial Patterns. *PLoS One*, **9** (3), e92097.

Kröncke I, Dippner J W, Heyen H and Zeiss B (1998). Long-Term Changes in Macrofaunal Communities off Norderney (East Frisia, Germany) in Relation to Climate Variability. *Marine Ecology Progress Series*, **167**, 25–36.

Kröncke I, Neumann H, Dippner J W, Holbrook S, Lamy T, Miller R, Padedda B M, Pulina S, Reed D C and Reinikainen M (2019). Comparison of Biological and Ecological Long-Term Trends Related to Northern Hemisphere Climate in Different Marine Ecosystems. *Nature Conservation*, **34**, 311–41.

Lindenmayer D, Burns E, Thurgate N and Lowe A (2014). The Value of Long-Term Research and How to Design Effective Ecological Research and Monitoring. In: Lindenmayer D, Burns E, Thurgate N and Lowe A, eds. *Biodiversity and Environmental Change: Monitoring, Challenges and Direction*, CSIRO, Collingwood, Vic., pp. 21–48.

Loreau M (2010). Linking Biodiversity and Ecosystems: Towards a Unifying Ecological Theory. *Philosophical Transactions of the Royal Society B*, **365**, 49–60.

Lotze H K, Lenihan H S, Bourque B J, Bradbury R H, Cooke R G, Kay M C, Kidwell S M, Kirby M X, Peterson C H and Jackson J B (2006). Depletion, Degradation, and Recovery Potential of Estuaries and Coastal Seas. *Science*, **312** (5781), 1806–9.

McCormick-Ray J (2005). Historical Oyster Reef Connections to Chesapeake Bay – a Framework for Consideration. *Estuarine, Coastal and Shelf Science*, **64** (1), 119–34.

McGeoch M A and Gaston K J (2002). Occupancy Frequency Distributions: Patterns, Artefacts and Mechanisms. *Biological Reviews*, **77**, 311–31.

Magurran A E and Henderson P A (2010). Temporal Turnover and the Maintenance of Diversity in Ecological

Assemblages. *Philosophical Transactions of the Royal Society B*, **365**, 3611–20.

Mahoney P and Bishop M (2017). Assessing Risk of Estuarine Ecosystem Collapse. *Ocean & Coastal Management*, **140**, 46–58.

Mearns A J, Reish D J, Oshida P S, Ginn T, Rempel-Hester M A, Arthur C, Rutherford N and Pryor R (2015). Effects of Pollution on Marine Organisms. *Water Environment Research*, **87** (10), 1718–816.

Ministry of Ecology and Environment (2018). *The 2017 Report on the State of the Ecology and Environment*, Ministry of Ecology and Environment, The People's Republic of China, Beijing.

Murphy G E and Romanuk T N (2012). A Meta-Analysis of Community Response Predictability to Anthropogenic Disturbances. *The American Naturalist*, **180** (3), 316–27.

Naeem S (1998). Species Redundancy and Ecosystem Reliability. *Conservation Biology*, **12**, 39–45.

Pacheco A S, Riascos J M, Orellana F and Oliva M E (2012). El Niño-Southern Oscillation Cyclical Modulation of Macrobenthic Community Structure in the Humboldt Current Ecosystem. *Oikos*, **121** (12), 2097–109.

Pascual M and Guichard F (2005). Criticality and Disturbance in Spatial Ecological Systems. *Trends in Ecology & Evolution*, **20** (2), 88–95.

Rietkerk M, Dekker S C, de Ruiter P C and van de Koppel J (2004). Self-Organized Patchiness and Catastrophic Shifts in Ecosystems. *Science*, **305** (5692), 1926–9.

Rodil I F, Lohrer A M, Hewitt J E, Townsend M, Thrush S F and Carbines M (2013). Tracking Environmental Stress Gradients Using Three Biotic Integrity Indices: Advantages of a Locally-Developed Traits-Based Approach. *Ecological Indicators*, **34**, 560–70.

Rousi H, Laine A O, Peltonen H, Kangas P, Andersin A-B, Rissanen J, Sandberg-Kilpi E and Bonsdorff E (2013). Long-Term Changes in Coastal Zoobenthos in the Northern Baltic Sea: The Role of Abiotic Environmental Factors. *ICES Journal of Marine Science: Journal du Conseil*, **70** (2), 440–51.

Schückel U and Kröncke I (2013). Temporal Changes in Intertidal Macrofauna Communities over Eight Decades: A Result of Eutrophication and Climate Change. *Estuarine, Coastal and Shelf Science*, **117**, 210–18.

Sellanesa J, Quirogac E, Neirad C and Gutierrez D (2007). Changes of Macrobenthos Composition under Different ENSO Cycle Conditions on the Continental Shelf off Central Chile. *Continental Shelf Research*, **27**, 1002–16.

Sheaves M (2009). Consequences of Ecological Connectivity: The Coastal Ecosystem Mosaic. *Marine Ecology Progress Series*, **391**, 107–15.

Takami H, Kawamura T, Won N I, Muraoka D, Hayakawa J and Onitsuka T (2017). Effects of Macroalgal Expansion Triggered by the 2011 Earthquake and Tsunami on Recruitment Density of Juvenile Abalone *Haliotis discus*

hannai at Oshika Peninsula, Northeastern Japan. *Fisheries Oceanography*, **26** (2), 141–54.

Taylor L R (1961). Aggregation, Variance, and the Mean. *Nature*, **189**, 732–5.

Thrush S F, Hewitt J E, Dayton P K, Coco G, Lohrer A M, Norkko A, Norkko J and Chiantore M (2009). Forecasting the Limits of Resilience: Integrating Empirical Research with Theory. *Proceedings of the Royal Society B*, **276**, 3209–17.

Thrush S F, Townsend M, Hewitt J E, Davies K F, Lohrer A M, Lundquist C and Cartner K (2013). The Many Uses and Values of Estuarine Ecosystems. In: Dymond J, ed. *Ecosystem Services in New Zealand—Condition and Trends*, Manaaki Whenua Press, Lincoln, New Zealand, pp. 226–37.

Tunberg B and Nelson W G (1998). Do Climatic Oscillations Influence Cyclical Patterns of Soft Bottom Macrobenthic Communities on the Swedish West Coast? *Marine Ecology Progress Series*, **170**, 85–94.

Turner E J and Miller D C (1991). Behaviour and Growth of *Mercenaria mercenaria* during Simulated Storm Events. *Marine Biology*, **111**, 55–64.

van Nes E H, Amaro T, Scheffer M and Duineveld G C A (2007). Possible Mechanisms for a Marine Benthic Regime Shift in the North Sea. *Marine Ecology Progress Series*, **330**, 39–47.

van Nes E H and Scheffer M (2005). Implications of Spatial Heterogeneity for Catastrophic Regime Shifts in Ecosystems. *Ecology*, **86**, 1797–807.

Ware D M and Thomson R E (2005). Bottom-up Ecosystem Trophic Dynamics Determine Fish Production in the Northeast Pacific. *Science*, **308**, 1280–4.

Warwick R M, Ashman C M, Brown A R, Clarke K R, Dowell B, Hart B, Lewis R E, Shillabeer N, Somerfield P J and Tapp J F (2002). Inter-Annual Changes in the Biodiversity and Community Structure of the Macrobenthos in Tees Bay and the Tees Estuary, UK, Associated with Local and Regional Environmental Events. *Marine Ecology Progress Series*, **234**, 1–13.

Warwick R M and Clarke K R (1993). Increased Variability as a Symptom of Stress in Marine Communities. *Journal of Experimental Marine Biology and Ecology*, **172**, 215–26.

Yachi S and Loreau M (1999). Biodiversity and Ecosystem Productivity in a Fluctuating Environment: The Insurance Hypothesis. *Proceedings of the National Academy of Sciences of the United States of America*, **96**, 1463–8.

Zajac R N, Vozarik J M and Gibbons B R (2013). Spatial and Temporal Patterns in Macrofaunal Diversity Components Relative to Sea Floor Landscape Structure. *PLoS One*, **8** (6), e65823.

Zampieri M, Toreti A, Schindler A, Scoccimarro E and Gualdi S (2017). Atlantic Multi-Decadal Oscillation Influence on Weather Regimes over Europe and the Mediterranean in Spring and Summer. *Global and Planetary Change*, **151**, 92–100.

PART IV

Functioning

CHAPTER 9

Ecosystem functions and the work of soft sediments

9.1 Introduction—the wide, wide world of functions

We often divide ecology into two components: structure and function (Figure 9.1). The structural elements include demographic descriptions of populations, species composition of communities and the habitat heterogeneity of ecosystems, all of which can vary in space and time. The functional element represents what these structural elements do, how they work together, and particularly how they interact with the environment to generate flows of energy and matter. Functioning involves biological, physical and chemical activities such as rates of primary and secondary production, carbon storage and nutrient cycling that may be unique for particular soft-sediment habitats in space and time. But functions are also used to define factors that influence ecosystem dynamics such as resilience or stability. Changes to the physical habitat or its biota, through natural disturbance, ecological succession or anthropogenic perturbation, will also change its functioning. Functioning is thus used as a measure of ecosystem health, productivity and contribution of the ecosystem to transformation, recycling and yield of energy and matter. The structural and functional elements of the ecosystem interact. For example, the effect of bioturbators on ecosystem functions, discussed in Chapter 2, changes in importance depending on the diversity and abundance of bioturbators, their levels of activity and the environment in which they are working.

Functions pervade our description of soft-sediment ecology and you will notice the concept is woven into many of the chapters. This is because to under-stand how ecosystems works is critically important, not only for the sake of scientific curiosity but also to allow us to manage our world—to weather climate change, to restore degraded ecosystems and to understand the past.

The widespread spatial dominance of marine soft sediments means that they play a pivotal role in overall marine ecosystem functioning (Snelgrove et al., 2014). Soft-sediment habitats produce multiple ecosystem functions that relate to nutrient and carbon cycling, sediment stability, creation of habitat, productivity and resilience and the animals and plants are key drivers for these functions. In shallow coastal environments the strong exchange between sediments and the water column means that sediments play a crucial role in regulating water column dynamics. The functional importance of marine sediments quickly becomes apparent when you think about the jargon phrase 'ecosystem engineers' (Jones et al., 1997). The idea of organisms engineering their own environment, influencing carbon and nutrient stocks, flows and interactions, relates in some way to every species that lives on or in soft sediments. Of course some engineers may be more important than others, but importance may vary depending on the function(s) being considered.

Thus you can see that ecological function is not a simple concept and it can be represented and studied in many ways. Regardless of how we study function, essentially we are trying to understand how the system works and how the system changes over time (i.e. its dynamics). Basically, we can think about three elements of function that relate to how it operates: its performance, persistence and port-

Ecology of Coastal Marine Sediments: Form, Function, and Change in the Anthropocene. Simon F. Thrush, Judi E. Hewitt, Conrad A. Pilditch and Alf Norkko, Oxford University Press (2021). © Simon Thrush, Judi Hewitt, Conrad Pilditch, and Alf Norkko.
DOI: 10.1093/oso/9780198804765.003.0009

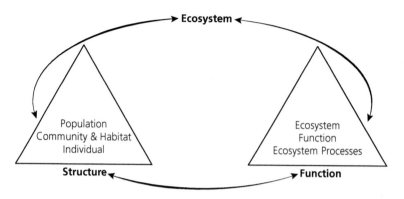

Figure 9.1 Ecology is comprised of two strongly linked components: structure and function.

Table 9.1 Elements of ecosystem function

Element	Relationship to delivery of the function(s)	Direct measurement
Performance	Quantity and speed or rate	Growth rates, demographics, biogeochemical fluxes
Persistence	Stability	Consistency in time
Portfolio	Multifunctionality	Simultaneous measurement of multiple functions

folio (Table 9.1). We will discuss these elements in this chapter.

9.2 Approaches to studying functions

We can think of ecosystem functions as the product of multiple ecosystem processes and the underpinning of multiple ecosystem services. Semantic confusion about ecosystem processes, functions and services arises when these terms are taken out of context. We define 'process' as the mechanism that underpins the function and 'service' as the emergent property that is valued by someone (Figure 9.2). A simple example is provided by the microphytobenthos (MPB; see Chapter 1). These microscopic plants secrete a mucus-like substance (EPS) that makes the sediment surface sticky (process); this stickiness can change the sediment erosion rate (function) affecting the stability of bedforms and water clarity (service). Not all connections are so straightforward, with interactions shifting in importance depending on the system and how we look at it. For example, habitat modification by animals through bioturbation of sediments is a process that supports ecosystem functions such as nutrient cycling, but also destabilisation of sediments.

9.2.1 Burrowing into rates and exchanges

Thinking about the units of measurement is a useful way of understanding what is being represented and how it might relate to other elements of the system you are studying. Some functions are represented by rates; for bioturbation this may be represented by units such as grams of sediment displaced per unit time per unit area (e.g. $g/h/m^2$). Functions may also be represented in terms of the outcome of the work, for example in terms of bioturbation we often describe it as the depth of sediment mixed with units of cm. Other functions are represented as quantities, for example habitat creation, or interaction webs—most commonly foodwebs, which are defined by measures of energy flow and web complexity (e.g. see Nordström & Bonsdorff, 2017).

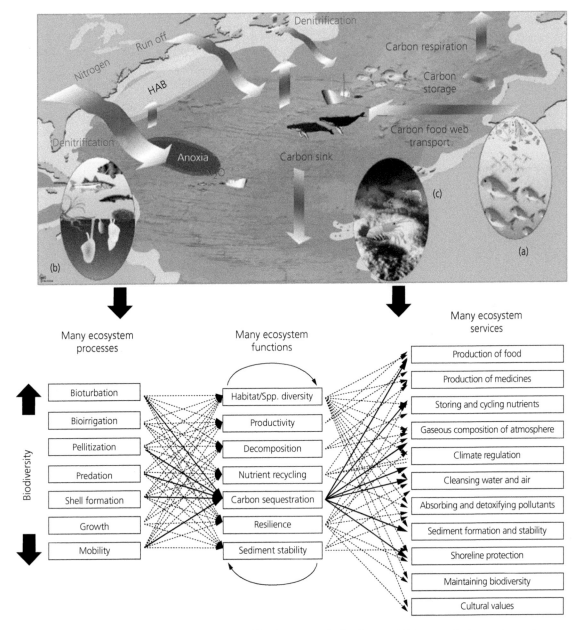

Figure 9.2 Multiple important processes structure ecological functions in coastal soft sediments, which in turn underpin ecosystem services. From Figure 1, Snelgrove et al. (2014). Reprinted from *Trends in Ecology & Evolution*, **29**, Snelgrove P V, Thrush S F, Wall D H, Norkko A, Real World Biodiversity-Ecosystem Functioning: A Seafloor Perspective, 398–405., Copyright (2014), with permission from Elsevier.

9.2.2 Functions and traits

The direct measurement of functions and processes is often time consuming, making broad-scale measurement difficult (see section 9.3). Biological or functional trait-based analyses of benthic commu-

nities (as discussed in Chapters 6 and 7) are a useful approach which links structure to function (Box 9.1). Traits can be representative of specific functions or aggregated into broader groups representing how specific groups of species are likely to respond to

---

Box 9.1 Generalising from traits to functions

There are many factors that make generalisation from traits to functions across studies difficult. Firstly, there is a diversity of interactions that link specific trait groups to functions, and there are variations in actual species composition within trait groups. Secondly, there can be important density-dependent or per-capita effects. Furthermore, animal–sediment interactions can result in functional plasticity across environmental gradients, resulting in habitat-dependent effects on functional traits, e.g. crabs burrowing in muddy or sandy sediments can have very different effects dependent on sediment cohesion (Needham et al., 2011). Finally, the strength of experimental contrasts between specific treatment and control conditions can vary. This emphasises the value of experimental studies that are conducted across multiple communities and habitats (Hillman et al., 2020a and b.

---

change (response traits) or how they drive functioning (effect traits) (Mensens et al., 2017; Suding et al., 2008). When traits are used to be representative of a specific function (e.g. nutrient cycling), morphological traits, such as size, may override structural aspects (biodiversity and composition), or abundance, as larger and older individuals can have disproportionately large effects on oxygen and nutrient fluxes (Norkko et al., 2013). Overall, the net direction of the inorganic nutrient flux due to bioturbation can also vary substantially with organism mobility, density or bioturbation mode. For example, surface-mixing amphipods can stimulate denitrification rates (Karlson et al., 2005), whereas deep-burrowing bioirrigating polychaetes sometimes have minimal effect on this process (Bonaglia et al., 2013).

9.2.3 Functions and systems approaches

Quantitative descriptions of ecosystem functioning were the cornerstone of *systems ecology* (Odum, 1964), that sought to define rates of carbon and nutrient fluxes, over space and time, through ecosystems. These systems approaches have been adapted into several different ecosystem- or earth-system models that currently are in use to understand carbon and nutrient flows. Whether derived

from systems ecology or geochemistry, these approaches often underpin the design of models used to describe and to some extent predict system performance (e.g. Ecopath and Ecosim to describe trophic relationships; Christensen & Walters, 2004) or other modelling approaches that incorporate sediments (Ehrnsten et al., 2020a). These 'system approaches' highlight a major challenge in finding a 'fit-for-purpose' balance between reductionist and holistic approaches to ecology (Bergandi & Blandin, 1998; Oreskes et al., 1994) as well as how the models are used and verified (Allison et al., 2018). In soft sediments, ecosystem functions and processes are at best currently poorly represented but often completely ignored (e.g. Ehrnsten et al., 2020b; Middelburg, 2018; Snelgrove et al., 2017).

9.3 Examples of key ecosystem functions in soft-sediment habitats

Here we discuss primary and secondary production and organic matter processing—nutrient cycling, especially for nitrogen, is discussed in Chapter 1. Examples of the overall effects of the process of bioturbation on a number of ecosystem functions are covered in Chapters 1, 2, 7 and 11.

9.3.1 Primary production

Soft-sediment habitats contain both autotrophic and heterotrophic organisms that generate biomass at the base of the foodweb. Primary production by MPB, seagrasses and macroalgae in marine sedimentary systems supports secondary production and other ecosystem functions, such as ecosystem metabolism, habitat formation, sediment stability and nutrient cycling.

Rates of primary production by sediment-dwelling microscopic algae, i.e. MPB, can be high, with estimates of 30–230 g carbon fixation m^{-2} $year^{-1}$ on intertidal flats (Heip et al., 1995; Hope et al., 2020). Globally it is estimated that MPB are responsible for the production of around 500 million tons of organic carbon annually (Cahoon, 1999). Moreover, in coastal photic areas, benthic primary productivity can far exceed pelagic productivity (Attard et al., 2019). Importantly, MPB production produces high-quality and readily assimilable food resources for consumers. However, macroalgae and vascular plants provide specific habitats as well as fulfilling their primary producer functions. Seagrass meadows have an important role in carbon sequestration (Macreadie et al., 2019), but their overall contribution to annual gross primary production (GPP) is more limited due to their limited spatial distribution. A seasonal comparison of major benthic primary producers at the habitat scale in the Finnish archipelago showed that macroalgae exhibited the highest annual GPP, which was twice as high as that for *Zostera marina* seagrass meadows (0.16 kg C m^{-2} year^{-1}; Figure 9.3; Attard et al., 2019). Beds of mixed macrophyte species and bare sediment with MPB, on the other hand, showed GPP approximately half that of seagrass meadows (0.10 and 0.07 kg C m^{-2} year^{-1}, respectively). Nevertheless, when accounting for the spatial extent in distribution of these key benthic primary producers, bare sediments were the most important (Figure 9.3).

Beyond the classical primary production by plants and algae that occurs through photosynthesis, with light as an energy source, there is also primary production that occurs through chemosynthesis. In chemosynthesis, the oxidation or reduction of inorganic chemicals is used to turn carbon-containing molecules (e.g. carbon dioxide or methane) into organic matter. Chemosynthesising bacteria are known to have symbiotic relationships with heterotrophs, and occur in both the deep sea and shallower sediments. Chemosynthesisers, particularly sulphur- and sulphide-oxidising and methane-oxidising (methanotrophic) bacteria, occur in a wide range of animal species (e.g. bivalves) that live in reducing environments, such as hydrothermal vents or low-oxygen and sulphide-rich environments. In dense seagrass beds, for example, there are lucinid bivalves that mostly live off carbon products provided by sulphide-oxidising chemoautotrophic bacteria (van der Geest et al., 2020). The recent discovery of cable bacteria that

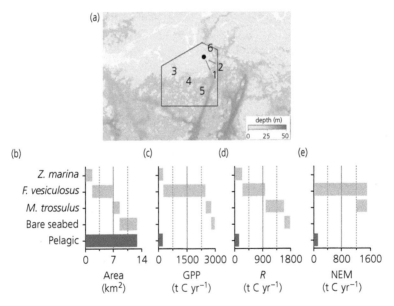

Figure 9.3 The distribution of shallow photic habitats (0–5 m) across *c.*13 km^2 in the northern Baltic Sea (close to Tvärminne Zoological Station) combined with spatial upscaling of annual metabolism rates for dominant benthic habitat types. Seagrasses on soft sediments (*Zostera marina*) and brown algae (*Fucus vesiculosus*) are dominant primary producers in terms of biomass, and mussel reefs dominated by Baltic blue mussels (*Mytilus trossulus*) are important secondary producers in the archipelago study area. The spatially most extensive habitats are bare soft sediments, which in terms of primary production are dominated by microphytobenthos. Habitat coverage (b) was combined with areal metabolism rates to estimate gross primary production (GPP, c), respiration (R, d) and net ecosystem metabolism (NEM, e). Pelagic estimates are from literature values and have been scaled linearly to water depth. GPP and R values are positive whereas NEM can be positive or negative. From Figure 4, Attard et al. (2019).

use centimetre-long filaments comprising thousands of cells in sediments (e.g. Bjerg et al., 2018; Burdorf et al., 2017) to connect sulphide oxidation with oxygen or nitrate reduction via long-distance electron transport, is likely to transform our understanding of the importance of chemosynthesis.

9.3.2 Secondary production

Secondary production is a key ecosystem function that reflects energy flow through the sediment ecosystems and via supporting higher trophic levels in the foodweb can export this material very widely (see Chapter 7). Production can be defined by the increase in biomass (organic matter), may be accumulated through growth as somatic production or as reproduction and is often conceptualised at the population or at a particular trophic level (e.g. detritivores, grazers, carnivores and parasites). Of course, some of the primary production consumed is lost through excretion and respiration. Secondary production is a key element of understanding energy flows and was a priority area before the 1980s, spurred by the interest in understanding the role of benthos for fish and fisheries (Holme & McIntyre, 1971; Warwick, 1980). Today we understand that there is more going on in marine sediments than the provision of fish food.

More recently, secondary production has emerged as a valuable indicator of the trophic capacity, health and functioning of aquatic ecosystems (Dolbeth et al., 2005). Biomass in itself is a fundamental organism trait that affects metabolic rates, energy demand and carbon uptake rates (Stachowicz et al., 2007). Ratios between production and biomass can be used to compare populations with different biomasses and turnover rates across habitats or locations (Elliott & Taylor, 1989; Emerson, 1989). Typically secondary productivity decreases with increasing size of macrofauna and production to biomass (P:B) ratios increase with a decrease in the size, longevity and age of fauna (Gray & Elliott, 2009). Estimating production as an ecosystem function is increasingly useful since many ecosystem services can be considered as proportional to increased biological production (e.g. expressed in energy equivalents as joules or as mg C m^{-2} d^{-1}) through the maintenance of significant carbon stocks of animals (Rodil et al., 2020).

9.3.3 Organic matter processing and ecosystem metabolism

Organic matter that is produced on or delivered to the seafloor directly fuels benthic foodwebs. Through the microbial processes of remineralisation, it also indirectly provides food resources and demineralises materials such as carbon and nutrients (see Chapter 1). Although mineralisation is primarily mediated by microbial processes including bacteria, and also meiofauna, macrofauna facilitate sediment organic carbon processing by microbes through consumption, the creation of geochemical microhabitats and moving sediment particles and porewater fluid (Glud, 2008). Some fauna, particularly suspension feeders, are involved in the process of bentho-pelagic coupling that can also significantly enhance the rates of organic matter deposition on the seafloor (Graf & Rosenberg, 1997) and in turn facilitate higher local abundances of deposit feeders that speed up organic matter mineralisation (Norkko et al., 2001).

All the metabolic processes that transform energy can be described by production and respiration. In benthic systems, the most common metrics include gross primary production (GPP) and respiration (R). Net ecosystem metabolism (NEM), i.e. the balance between GPP and R, is a useful indicator of ecosystem-level trophic conditions. NEM can in itself therefore be used as a functionality metric for contrasting benthic communities and habitats across time and space in terms of their overall productivity and the relative balance between auto- and heterotrophy, similar to metrics used in biological oceanography (Attard et al., 2019; Rodil et al., 2020).

The remineralisation of organic matter increases the respiration of soft sediments and consumes oxygen, increasing benthic oxygen consumption. The release of primary nutrients as a consequence of remineralisation can enhance primary production in non-eutrophic sediments. This emphasises the importance of considering the time scales of different processes that contribute to functions and the need to remember that when our measurements are net effects these are composed of the balance between positive and negative processes. Although bacterial respiration dominates (Arndt et al., 2013), studies of oxygen consumption have shown that

Figure 9.4 Habitat structure is enhanced in soft sediments both above and below the sediment surface. Mollusc shells can dominate sediments and provide settlement surfaces for the establishment of epifauna. Sponges create structures that provide new habitats and influence predator density. Burrowing crustaceans create deep and often complex, connecting below-ground structures. (Photo credits Simon Thrush and Robotics Lab, CNRS, Genoa, Italy.)

macrofauna can contribute substantially to total sediment oxygen uptake, either directly through respiration or indirectly through particle and solute mixing, i.e. through bioturbation (Glud, 2008). Respiration is strongly dependent on abundance and particularly on the biomass of macrofaunal communities and their dominant functional groups (Janson et al., 2012, 2013; Norkko et al., 2013). Global estimates are highly variable, but macrofaunal respiration has been estimated to account for up to 45–70% of benthic carbon mineralisation in coastal areas dominated by, e.g. mussels (Rodil et al., 2020; Wenzhofer & Glud, 2004), whereas Herman et al. (1999) estimated median values of 15–20% for shallow estuaries.

9.3.4 Physical structuring of sedimentary habitats by marine organisms

Biogenic habitat structuring by foundation species (Dayton, 1979), both above and below the sediment surface, affects multiple functions (Thrush & Dayton, 2002). Increases in the structural complexity of habitat increase habitat diversity and the potential for co-existence of multiple species and have major implications for sediment stability and recruitment dynamics of benthic macrofauna (Friedrichs et al., 2000; Thrush et al., 1996). Many of the organisms creating structure above the surface are calcifying organisms such as gastropods and bivalves, that can make a significant contribution to carbonate sediment generation (Smith et al., 2010). Bivalve reefs, seagrass or sponge beds all influence benthic boundary flow dynamics (Chapter 3), influencing the exchange of inorganic and organic particles (Wildish & Kristmanson, 1997). Habitat structuring effects obviously also extend into the seafloor, as plants and animals extend their roots, burrows or tube structures below the sediment surface with major effects on sediment stability and sediment oxygenation (Figure 9.4).

9.4 Resilience, functional redundancy— and stability

There is another suite of ecosystem functions that do not directly link organisms to fluxes of energy and matter, but rather support the persistence of ecosystems in the face of change and thus represent resilience and stability. Resilience is an emergent property of complex systems and relates to the adaptive capacity of the ecosystem to respond to change. There are limits to resilience and when these are exceeded ecosystems can shift into alternative states (see also Chapter 8). The functions that link to resilience relate to those that modify the impact of stressors and cumulative effects and those that confer recovery from disturbance and sustain functional performance through functional redundancy (see Chapter 10). Understanding resilience involves understanding interactions across scales, from the individual species (e.g. biological species traits) to community (e.g. functional redundancy) and landscape scales (e.g. species occupancy across the landscape). Thus, resilience links concepts across scales of biological organisation (Gladstone-Gallagher et al., 2019).

The plethora of functions provided by marine soft sediments are intimately linked with biodiversity, i.e. higher biodiversity entails a wider range of functional traits that provides for more efficient resource use, including as well as providing stability to ecosystem functioning in variable environments and in the face of disturbance (Chapin III et al., 1997). The roles of trait diversity and redundancy in benthic communities (see Chapter 10) are all linked to the capacity of the system to cope with change. For example, the relationship between productivity, eutrophication and organic matter remineralisation involves organisms living in sediments transforming organic material, releasing of nutrients, supporting recycling, and transfer through foodwebs (Bourgeois et al., 2017; Villnäs et al., 2012). Non-eutrophic sediments with diverse ecological communities are able to process and remove large quantities of nitrogen and frequently bind phosphorus. As sediment nutrient loads increase and species contributing to ecosystem functioning decrease, increasing stress is likely to result in non-linear change or tipping points in the ecosystem's ability to cope with nutrient load (Folke et al., 2004).

9.5 Functional dynamics—connections and patches

Functionality is all about connections, with a major research focus on understanding how functions vary in space and time. This understanding is vital for our ability to develop environmental future scenarios. For instance, we may measure nutrient processing at scales of millimetres to centimetres associated with microbial processes, but these processes are likely influenced by bioturbation by macrofauna on the scale of metres and may be further influenced by predator disturbance of the sediment on scales of tens to hundreds of metres. Carbon may be converted into complex organic molecules and stored in MPB for hours to days before it is eaten by a shellfish that may store some of the carbon for decades or be consumed by fish, birds or mammals. Less assimilable plants such as seagrass and mangroves may store carbon in their tissues for much longer and degradation pathways may be slower. Part of the assimilation and breakdown processes may involve the subduction of carbon deep into the sediment where it can be sequestered below the bioturbation zone for decades to centuries. Carbon that is trophically transferred into large animals that die and sink in the deep sea may be effectively sequestered for centuries to millennia. These interactions of processes and functions working across different space and time scales lead to complex system dynamics and make functions (and ecosystem services—see section 9.7) emergent properties of ecosystems.

The nature, form and diversity of soft-sediment ecosystem functions can be profoundly changed by human activities at multiple scales (see Chapter 11). Toxic contaminants generally alter marine ecosystem functioning by reducing productivity and increasing respiration (Johnston et al., 2015). A fascinating study of century-scale change on the continental shelf of southern California provides a good example based on a geological approach to recent ecological transitions (Tomašových & Kidwell, 2017). Abundant old shells of suspension-feeding brachiopods and scallops are common on the shelf, but the shelf is now dominated by muddy sediments. The loss of the previous suspension-feeder community is linked to human activity on land associated with land conversion and the concomitant increased sediment deposition and siltation on the shelf in the nineteenth century. The loss of suspension feeders, combined with the physical disturbance and siltation, likely changed multiple ecosystem functions associated with reduced habitat heterogeneity, bentho-pelagic coupling, shell production and sediment biogeochemistry.

9.5.1 Capturing rapid dynamics

Molecular techniques are showing potential as a way to enhance our observation of ecological change (Dafforn et al., 2014). 'Omic' approaches help elucidate what microbial processes are working at specific places and moments of time, revealing the underlying genetic mechanisms of observed altered functions (Birrer et al., 2017, 2019). Assays of extracellular enzyme activity in marine sediments are also a useful tool to help inform our understanding of ecosystem processes and functions (Crawshaw et al., 2019). These techniques will undoubtedly play an increasing role in our understanding of the dynamics of ecosystem functions, but they, as with any data collection method, need to have clear sampling strategies that account for the spatial and temporal scales of the processes being examined (see Graur et al., 2015 and Chapters 2, 4 and 5).

9.5.2 Capturing broad-scale spatial dynamics in ecosystem functions

Developing a framework to quantify the transformation, sequestration or connectivity of functions across habitat patches in ways that are sensitive to ecological change is a major challenge for ecology (Hillman et al., 2018). This requires defining the relationships between functions (performance, persistence and portfolio) and habitat type to understand how changes in the connectivity between habitats influence import, export, recycling, transformation and sequestration of energy and matter. Research on the interconnectedness of habitats and ecosystems builds on work on spatial subsidies and source–sink dynamics in foodwebs (Polis et al., 2004). Gillis et al. (2014) identified three basic requirements for positive functional connections across habitats: (1) the quantity and quality of resource exchange from a donor are large enough to elicit a response in the receiving habitat, which links to habitat proximity and connectivity; (2) material import or export must benefit the donor or receiving habitats, which relates to resource requirements or stress-buffering capacity; and (3) these exchanges involve functionally important species that drive landscape-scale facilitation. To understand change in these large-scale functional dynamics we need spatial mapping tools that identify how functional hotspots shift in location and intensity under different scenarios. For example, Hillman et al. (2020b) highlighted the role of different estuarine habitats in affecting the trajectory of change in ecosystem function, using Zonation as a spatial mapping tool. The study focussed on a large sediment-dwelling animal whose behaviour has important influences on ecosystem functions and which suffers declines in abundance with sediment deposition.

9.6 Multifunctionality

Variations in space and time and the difficulties of empirical measurements have meant that much of the research on ecosystem function in marine sediments has tended to take a single-function focus, for example examining productivity or carbon sequestration. We need to remember that the functionality of marine sediments is multidimensional—the sediments perform multiple functions. This collective output is called multifunctionality (Byrnes et al., 2014; Hector & Bagchi, 2007).

Measuring multiple functions in marine sediments is possible, particularly when we are concerned with similar kinds of processes. For example, benthic chambers can be used to simultaneously assess oxygen consumption (an indicator of remineralisation), primary production, nutrient flux and denitrification. All of these functions are intimately linked to biogeochemical processes and so we can expect to infer relationships between the individual response variables (Lohrer et al., 2011). When we want to include other types of function, e.g. sediment creation, stabilisation, secondary production, resilience or habitat creation, we can use a suite of measurements or proxies of functions that are/represent rate measurements (e.g. Villnäs et al., 2013). Nevertheless, we need to think about both the appropriate scales of measurement and the interconnectedness of processes. Evolving techniques and rapid assessment processes will help to address this challenge in the future, which means that we will not be required to overly simplify ecosystem complexity. However, at present measures of ecosystem functioning that account for multifunctionality are rare.

One solution to the current difficulties of empirical measurement of multifunctionality is the use of ecological traits as surrogates for function (Figure 9.5). This kind of approach recognises that individual traits rarely occur randomly in species, but rather as smaller subsets of trait groupings (in terms of size, longevity, dispersal mode etc.) that relate strongly to specific benthic ecosystem properties and functions (e.g. bioturbation, community stability and juvenile dispersal potential). The functional trait subgroup may make a unique contribution to a cumulative expression of an ecosystem property and an additive analysis of the combined functional trait groupings allows for a spatial illustration of benthic ecosystem multifunctionality. The benefit is that data on the structure of benthic communities can be interpreted in terms of proxies for functionality across broad-scale geographic regions and environmental gradients with quantitative estimates of multifunctionality. This can be very useful in a marine spatial planning and management perspective, but requires both expert insight regarding the natural history of species and subsequent weighting of traits as well as large-scale data.

Assigning traits to specific species requires some level of expert knowledge and this knowledge can also be used to define links between traits, processes and multiple functions. This alternative route to studying multifunctionality in complex environmental systems provides for a transparent analysis of the links between multiple ecosystem components in the provision of ecosystem functions as well as demonstrating the foundational role of ecosystem functions in delivering ecosystem services.

9.7 From ecosystem function to ecosystem services

Ecosystem services represent a bridge between ecosystem functions and human values. The concept of ecosystem services is a way of introducing to society ideas about how the often unseen and unvalued species and processes that occur in marine sediments directly or indirectly add value to humanity (Daily et al., 2011). For example, estimates indicate that at least 80% of terrestrial dissolved inorganic nitrogen can be denitrified in the coastal ocean margin (Erisman et al., 2013).

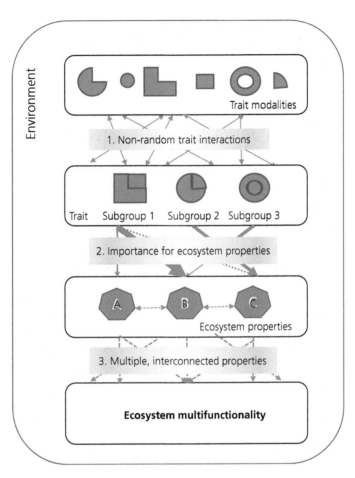

Figure 9.5 Conceptual figure showing how non-random trait interactions, of varying importance, can result in (spatial) patterns in ecosystem multifunctionality. From Figure 2, Villnäs et al. (2018).

We can think of ecosystem services as linking knowledge of ecosystems, society and economies, but here we are focussed on the foundational relationship between function and services. Ecosystem services have been categorised in many ways, with four classes commonly considered: provisioning services relate to production functions; regulation and maintenance services relate to functions associated with biogeochemistry, sediment stability and resilience; habitat services relate to functions that create and maintain biogenic structure; and cultural services that while dependent on different knowledge and value systems, link to all functions that provide for a sense of wildness, place, well-being and learning. These services are the infrastructure that supports the marine economy and are vital to maintain the benefits and services we derive from the oceans.

We have already highlighted the complexity of links between ecosystem processes and functions and the need to consider how ecosystems work on different space and time scales. Moving to consider services from an ecological perspective does nothing to reduce this complexity (see Figure 9.2). This means that there is a high potential for unintended consequences if we focus solely on the services for which there is a marketable value—such as food production.

Just as there are multiple functions operating in soft sediments, so too are there multiple services. Coastal marine ecosystems such as harbours and estuaries not only provide goods and services for use within estuaries but also disproportionately influence the nature and rate of biogeochemical processes that sustain the biosphere. These shallow, comparatively warm, sunlit, well-mixed waters and extensive soft-sediment habitats play significant roles in processing contaminants from land and fuelling productivity on the adjacent coast. Fish live within and pass through estuaries, either to spawn in rivers or to spend their adult life in the open sea. This focus on the role of connectivity of functions in contributing to services needs to be carefully defined if we are to optimise marine spatial management benefits (Foley et al., 2010; Van der Biest et al., 2020).

Defining, mapping and assessing change in ecosystem services are especially important to soft-sediment ecology because all our fundamental functional research points to the importance of these ecosystems. Despite this, the diversity of marine ecosystem services is usually poorly characterised in ecosystem service assessments, because the critical scale of heterogeneity in the sediments does not always map to easily observable habitat features. As a consequence, studies have tended to focus on easy-to-see habitats like mangroves, seagrass and oyster reefs—all of which are important, but so is the activity in many other soft-sediment habitats.

The development of ecosystem services for marine systems has highlighted multiple data limitations. Possibly the easiest to understand, but difficult to solve, is measuring and modelling the connectivity between where a service is produced and where it is used or valued. This can represent a critical mismatch in the locations of ecosystem functions and services. For example, shallow-sediment habitats dominated by shellfish beds or seagrass can provide critical nursery habitat for fish. This function can be generally located well away from areas where fish are caught and people experience the joys (or otherwise, depending on luck) of recreational fishing.

9.8 Close out

Importantly, animals and plants are really the powerhouses that modify transformation rates of carbon and nutrients. In soft-sediment systems,

they are also often key drivers of biodiversity, resilience and recovery. Overall, they are key to maintaining ecosystem functions and services (see Figure 2.5 in Chapter 2). With increasing human dominance of marine ecosystems, and rapidly changing nutrient and carbon cycles, biodiversity and habitat provision, understanding and quantifying ecosystem functions and the mechanisms underpinning them are becoming increasingly important.

Ultimately ecological value depends on the quantity of intact ecosystem functions (Limburg, 1999). Unfortunately, perspectives on values, states and trends are easily biased by shifting baselines that plague ecological comparisons when information on ecosystem history is limited (Dayton et al., 1998; Duarte et al., 2009). This problem is often confounded for soft-sediment systems which lie largely out of sight. The concept of ecosystem services is thus more critical to soft-sediment habitats than to others. It provides a way to demonstrate the value to society of soft-sediment habitats that are often ignored and undervalued in environmental management decision making (Granek et al., 2009; Davies et al., 2015).

Fundamental crossover social and bio-physical science research is required (Bastow et al., 2014) to better (1) understand this complex new marine world of shifting resource use, ecosystem function, biodiversity and environmental change; (2) identify and respond to the likely consequences of decisions made in it; (3) communicate this knowledge to society; and (4) engage different groups in making constrained choices and developing compromise. Integrated studies will hopefully generate robust tools for defining ecosystem benefits and provide communities with better ways to make choices and trade-offs about how resources are managed (Biggs et al., 2012).

References

Allison A E F, Dickson M E, Fisher K T and Thrush S F (2018). Dilemmas of Modelling and Decision-Making in Environmental Research. *Environmental Modelling and Software*, **99**, 147–55.

Arndt S, Jørgensen B B, LaRowe D E, Middelburg J J, Pancost R D and Regnier P (2013). Quantifying the

Degradation of Organic Matter in Marine Sediments: A Review and Synthesis. *Earth-Science Reviews*, **123**, 53–86.

Attard K M, Rodil I F, Glud R N, Berg P, Norkko J and Norkko A (2019). Seasonal Ecosystem Metabolism across Shallow Benthic Habitats Measured by Aquatic Eddy Covariance. *Limnology and Oceanography Letters*, **4**, 79–86.

Bastow S, Dunleavy P and Tinkler J (2014). *The Impact of the Social Sciences: How Academics and Their Research Make a Difference*, Sage, London.

Bergandi D and Blandin P (1998). Holism vs. Reductionism: Do Ecosystem Ecology and Landscape Ecology Clarify the Debate? *Acta Biotheoretica*, **46** (3), 185–206.

Biggs R, Schlüter M, Biggs D, Bohensky E L, Burnsilver S, Cundill G, Dakos V, Daw T M, Evans L S, Kotschy K, Leitch A M, Meek C, Quinlan A, Raudsepp-Hearne C, Robards M D, Schoon M L, Schultz L and West P C (2012). Toward Principles for Enhancing the Resilience of Ecosystem Services. *Annual Review of Environment and Resources*, **37**, 421–48.

Birrer S C, Dafforn K A and Johnston E L (2017). Microbial Community Responses to Contaminants and the Use of Molecular Techniques. In: Cravo-Laureau C, Cagnon C, Duran R and Lauga B, eds. *Microbial Ecotoxicology*, Springer International, Cham, pp. 165–83.

Birrer S C, Dafforn K A, Sun M Y, Williams R B H, Potts J, Scanes P, Kelaher B P, Simpson S L, Kjelleberg S, Swarup S, Steinberg P and Johnston E L (2019). Using Meta-Omics of Contaminated Sediments to Monitor Changes in Pathways Relevant to Climate Regulation. *Environmental Microbiology*, **21** (1), 389–401.

Bjerg J T, Boschker H T S, Larsen S, Berry D, Schmid M, Millo D, Tataru P, Meysman F J R, Wagner M, Nielsen L P and Schramm A (2018). Long-Distance Electron Transport in Individual, Living Cable Bacteria. *Proceedings of the National Academy of Sciences of the United States of America*, **115** (22), 5786–91.

Bonaglia S, Bartoli M, Gunnarsson J S, Rahm L, Raymond C, Svensson O, Yekta S S and Brüchert V (2013). Effect of Reoxygenation and *Marenzelleria* Spp. Bioturbation on Baltic Sea Sediment Metabolism. *Marine Ecology Progress Series*, **482**, 43–55.

Bourgeois S, Archambault P and Witte U (2017). Organic Matter Remineralization in Marine Sediments: A Pan-Arctic Synthesis. *Global Biogeochemical Cycles*, **31** (1), 190–213.

Burdorf L D W, Tramper A, Seitaj D, Meire L, Hidalgo-Martinez S, Zetsche E M, Boschker H T S and Meysman F J R (2017). Long-Distance Electron Transport Occurs Globally in Marine Sediments. *Biogeosciences*, **14** (3), 683–701.

Byrnes J E K, Gamfeldt L, Isbell F, Lefcheck J S, Griffin J N, Hector A, Cardinale B J, Hooper D U, Dee L E and Emmett Duffy J (2014). Investigating the Relationship between Biodiversity and Ecosystem Multifunctionality: Challenges and Solutions. *Methods in Ecology and Evolution*, **5** (2), 111–24.

Cahoon L B (1999). The Role of Benthic Microalgae in Neritic Ecosystems. *Oceanography and Marine Biology*, **37**, 47–86.

Chapin III F S, Walker B H, Hobbs R J, Hooper D U, Lawton J H, Sala O E and Tilman D (1997). Biotic Control over the Functioning of Ecosystems. *Science*, **277** (5325), 500–4.

Christensen V and Walters C J (2004). Ecopath with Ecosim: Methods, Capabilities and Limitations. *Ecological Modelling*, **172**, 109–39.

Crawshaw J, O'Meara T, Savage C, Thomson B, Baltar F and Thrush S F (2019). Source of Organic Detritus and Bivalve Biomass Influences Nitrogen Cycling and Extracellular Enzyme Activity in Estuary Sediments. *Biogeochemistry*, **145** (3), 315–35.

Dafforn K A, Baird D J, Chariton A A, Sun M Y, Brown M V, Simpson S L, Kelaher B P and Johnston E L (2014). Faster, Higher and Stronger? The Pros and Cons of Molecular Faunal Data for Assessing Ecosystem Condition. *Advances in Ecological Research*, **51**, 1–40.

Daily G C, Kareiva P, Polasky S, Ricketts T H and Tallis H (2011). Mainstreaming Natural Capital into Decisions. In: Kareiva P, Tallis H, Ricketts T H, Daily G C and Polasky S, eds. *Natural Capital: Theory and Practice of Mapping Ecosystem Services*, Oxford University Press, Oxford, UK, pp. 3–14.

Davies K K, Fisher K T, Dickson M E, Thrush S F and Le Heron R (2015). Improving Ecosystem Service Frameworks to Address Wicked Problems. *Ecology and Society*, **20** (2), 37.

Dayton P K (1979). Observations on Growth Dispersal and Population Dynamics of Some Sponges in McMurdo Sound, Antarctica. In: Levi C and Boury-Esnault N, eds. *Biologie De Spongiares*, Colloque International CNRS, Paris, pp. 271–82.

Dayton P K, Tegner M J, Edwards P B and Riser K L (1998). Sliding Baselines, Ghosts, and Reduced Expectations in Kelp Forest Communities. *Ecological Applications*, **8** (2), 309–22.

Dolbeth M, Lillebø A I, Cardoso P G, Ferreira S M and Pardal M A (2005). Annual Production of Estuarine Fauna in Different Environmental Conditions: An Evaluation of the Estimation Methods. *Journal of Experimental Marine Biology and Ecology*, **326** (2), 115–27.

Duarte C M, Conley D, Carstensen J and Sanchez-Comacho M (2009). Return to Neverland: Shifting Baselines Affect Eutrophication Restoration Targets. *Estuaries and Coasts*, **32**, 29–36.

Ehrnsten E, Norkko A, Müller-Karulis B, Gustafsson E and Gustafsson B G (2020b). The Meagre Future of Benthic

Fauna in a Coastal Sea—Benthic Responses to Recovery from Eutrophication in a Changing Climate. *Global Change Biology*, **26** (4), 2235–50.

Ehrnsten E, Sun X, Humborg C, Norkko A, Savchuk O, Slomp C P, Timmermann K and Gustafsson B G (2020a). Understanding Environmental Changes in Temperate Coastal Seas: Linking Models of Benthic Fauna to Carbon and Nutrient Fluxes. *Frontiers in Marine Science*, **7**, 450.

Elliott M and Taylor C J L (1989). The Production Ecology of the Subtidal Benthos of the Forth Estuary, Scotland. *Scientia Marina*, **53**, 531–41.

Emerson C W (1989). Wind Stress Limitation of Benthic Secondary Production in Shallow, Soft Sediment Communities. *Marine Ecology Progress Series*, **53**, 65–77.

Erisman J W, Galloway J N, Seitzinger S, Bleeker A, Dise N B, Petrescu A M, Leach A M and de Vries W (2013). Consequences of Human Modification of the Global Nitrogen Cycle. *Philosophical Transactions of the Royal Society of London. Series B, Biological Sciences*, **368** (1621), 2013.0116.

Foley M M, Halpern B S, Micheli F, Armsby M H, Caldwell M R, Crain C M, Prahler E, Rohr N, Sivas D, Beck M W, Carr M H, Crowder L B, Emmett Duffy J, Hacker S D, McLeod K L, Palumbi S R, Peterson C H, Regan H M, Ruckelshaus M H, Sandifer P A and Steneck R S (2010). Guiding Ecological Principles for Marine Spatial Planning. *Marine Policy*, **34** (5), 955–66.

Folke C, Carpenter S, Walker B, Scheffer M, Elmqvist T, Gunderson L and Holling C S (2004). Regime Shifts, Resilience and Biodiversity in Ecosystem Management. *Annual Review of Ecology and Systematics*, **35**, 557–81.

Friedrichs M, Graf G and Springer B (2000). Skimming Flow Induced over a Simulated Polychaete Tube Lawn at Low Population Densities. *Marine Ecology Progress Series*, **192**, 219–28.

Gillis L G, Bouma T J, Jones C G, Van Katwijk M M, Nagelkerken I, Jeuken C J L, Herman P M J and Ziegler A D (2014). Potential for Landscape-Scale Positive Interactions among Tropical Marine Ecosystems. *Marine Ecology Progress Series*, **503**, 289–303.

Gladstone-Gallagher R V, Pilditch C A, Stephenson F and Thrush S F (2019). Linking Traits across Ecological Scales Determines Functional Resilience. *Trends in Ecology and Evolution*, **34** (12), 1080–91.

Glud R N (2008). Oxygen Dynamics of Marine Sediments. *Marine Biology Research*, **4** (4), 243–89.

Graf G and Rosenberg R (1997). Bioresuspension and Biodeposition: A Review. *Journal of Marine Systems*, **11** (3–4), 269–78.

Granek E F, Polasky S, Kappel C V, Reed D J, Stoms D M, Koch E W, Kennedy C J, Cramer L A, Hacker S D, Barbier E B, Aswani S, Ruckelshaus M H, Perillo G M E,

Silliman B R, Muthiga N, Bael D and Wolanski E (2009). Ecosystem Services as a Common Language for Coastal Ecosystem-Based Management. *Conservation Biology*, **24** (1), 207–16.

Graur D, Zheng Y and Azevedo R B R (2015). An Evolutionary Classification of Genomic Function. *Genome Biology and Evolution*, **7** (3), 642–5.

Gray J S and Elliott M (2009). *Ecology of Marine Sediments: From Science to Management*, 2nd ed., Oxford University Press, Oxford, UK.

Hector A and Bagchi R (2007). Biodiversity and Ecosystem Multifunctionality. *Nature*, **448**, 188–91.

Heip C H R, Goosen N K, Herman P M J, Kromkamp J, Middelburg J J and Soetaert K (1995). Production and Consumption of Biological Particles in Temperate Tidal Estuaries. *Oceanography and Marine Biology: An Annual Review*, **33**, 1–149.

Herman P M J, Middelburg J J, Van de Koppel J and Heip C H R (1999). Ecology of Estuarine Macrobenthos. *Advances in Ecological Research*, **29**, 195–231.

Hillman J R, Lundquist C, O'Meara T and Thrush S (2020a). Large Animal Depletions Influence Nutrient Processes Fluxes in a Heterogeneous Marine Intertidal Soft-Sediment Ecosystem. *Ecosystems*.

Hillman J R, Lundquist C J, Pilditch C A and Thrush S F (2020b). The Role of Large Macrofauna in Mediating Sediment Erodibility across Sedimentary Habitats. *Limnology and Oceanography*, **65** (4), 683–93.

Hillman J R, Lundquist C J and Thrush S F (2018). The Challenges Associated with Connectivity in Ecosystem Processes. *Frontiers in Marine Science*, **5**, 364.

Hillman J R, Stephenson F, Thrush S F and Lundquist C J (2020b). Investigating Changes in Estuarine Ecosystem Functioning under Future Scenarios. *Ecological Applications*, **30** (4), e02090.

Holme N A and McIntyre A D (1971). *Methods for Studying Marine Benthos*, IBP Handbook No. 16, Blackwell Scientific, Oxford, UK.

Hope J A, Paterson D M and Thrush S (2020). The Role of Microphytobenthos in Soft-Sediment Ecological Networks and Their Contribution to the Delivery of Multiple Ecosystem Services. *Journal of Ecology*, **108** (3), 815–30.

Janson A L, Denis L, Rauch M and Desroy N (2012). Macrobenthic Biodiversity and Oxygen Uptake in Estuarine Systems: The Example of the Seine Estuary. *Journal of Soils and Sediments*, **12** (10), 1568–80.

Janson A L, Denis L, Rauch M and Desroy N (2013). Erratum to Macrobenthic Biodiversity and Oxygen Uptake in Estuarine Systems: The Example of the Seine Estuary (J Soils Sediments, (2012), 12, (1568–80), 10.1007/S11368-012-0557-2). *Journal of Soils and Sediments*, **13** (4), 834–5.·

Johnston E L, Mayer-Pinto M and Crowe T P (2015). Chemical Contaminant Effects on Marine Ecosystem Functioning. *Journal of Applied Ecology*, **52** (1), 140–9.

Jones C G, Lawton J H and Shachak M (1997). Positive and Negative Effects of Organisms as Physical Ecosystem Engineers. *Ecology*, **78** (7), 1946–57.

Karlson K, Hulth S, Ringdahl K and Rosenberg R (2005). Experimental Recolonisation of Baltic Sea Reduced Sediments: Survival of Benthic Macrofauna and Effects on Nutrient Cycling. *Marine Ecology Progress Series*, **294**, 35–49.

Limburg K E (1999). Estuaries, Ecology, and Economic Decisions: An Example of Perceptual Barriers and Challenges to Understanding. *Ecological Economics*, **30** (1), 185–8.

Lohrer A M, Hewitt J E, Hailes S F, Thrush S F, Ahrens M and Halliday J (2011). Contamination on Sandflats and the Decoupling of Linked Ecological Functions. *Austral Ecology*, **36** (4), 378–88.

Macreadie P I, Anton A, Raven J A, Beaumont N, Connolly R M, Friess D A, Kelleway J J, Kennedy H, Kuwae T, Lavery P S, Lovelock C E, Smale D A, Apostolaki E T, Atwood T B, Baldock J, Bianchi T S, Chmura G L, Eyre B D, Fourqurean J W, Hall-Spencer J M, Huxham M, Hendriks I E, Krause-Jensen D, Laffoley D, Luisetti T, Marbà N, Masque P, McGlathery K J, Megonigal J P, Murdiyarso D, Russell B D, Santos R, Serrano O, Silliman B R, Watanabe K and Duarte C M (2019). The Future of Blue Carbon Science. *Nature Communications*, **10** (1), 3998.

Mensens C, De Laender F, Janssen C R, Sabbe K and De Troch M (2017). Different Response–Effect Trait Relationships Underlie Contrasting Responses to Two Chemical Stressors. *Journal of Ecology*, **105** (6), 1598–609.

Middelburg J J (2018). Reviews and Syntheses: To the Bottom of Carbon Processing at the Seafloor. *Biogeosciences*, **15** (2), 413–27.

Needham H R, Pilditch C A, Lohrer A M and Thrush S F (2011). Context-Specific Bioturbation Mediates Changes to Ecosystem Functioning. *Ecosystems*, **14**, 1096–109.

Nordström M C and Bonsdorff E (2017). Organic Enrichment Simplifies Marine Benthic Food Web Structure. *Limnology and Oceanography*, **62** (5), 2179–88.

Norkko A, Hewitt J E, Thrush S F and Funnell G A (2001). Benthic-Pelagic Coupling and Suspension Feeding Bivalves: Linking Site-Specific Sediment Flux and Biodeposition to Benthic Community Structure. *Limnology and Oceanography*, **46**, 2067–72.

Norkko A, Villnäs A, Norkko J, Valanko S and Pilditch C (2013). Size Matters: Implications of the Loss of Large Individuals for Ecosystem Function. *Scientific Reports*, **3**, 2646.

Odum E P (1964). The New Ecology. *Bioscience*, **14**, 14–16.

Oreskes N, Shrader-Frechette K and Belitz K (1994). Verification, Validation, and Confirmation of Numerical Models in the Earth Sciences. *Science*, **263** (5147), 641–6.

Polis G A, Power M E and Huxel G R (2004). *Food Webs at the Landscape Level*, University of Chicago Press, Chicago, IL.

Rodil I F, Attard K M, Norkko J, Glud R N and Norkko A (2020). Estimating Respiration Rates and Secondary Production of Macrobenthic Communities across Coastal Habitats with Contrasting Structural Biodiversity. *Ecosystems*, **23** (3), 630–47.

Smith A M, Wood A C L, Liddy M F A, Shears A E and Fraser C I (2010). Human Impacts in an Urban Port: The Carbonate Budget, Otago Harbour, New Zealand. *Estuarine, Coastal and Shelf Science*, **90** (2), 73–9.

Snelgrove P V R, Soetaert K, Solan M, Thrush S, Wei C-L, Danovaro R, Fulweiler R W, Kitazato H, Ingole B, Norkko A, Parkes R J and Volkenborn N (2017). Global Carbon Cycling on a Heterogeneous Seafloor. *Trends in Ecology & Evolution*, **33** (2), 96–105.

Snelgrove P V R, Thrush S F, Wall D H and Norkko A (2014). Real World Biodiversity–Ecosystem Functioning: A Seafloor Perspective. *Trends in Ecology & Evolution*, **29** (7), 398–405.

Stachowicz J J, Bruno J F and Duffy J E (2007). Understanding the Effects of Marine Biodiversity on Communities and Ecosystems. *Annual Review of Ecology and Systematics*, **38**, 739–66.

Suding K N, Lavorel S, Chapin F S, Cornelissen J H C, Diaz S, Garnier E, Goldberg D, Hooper D U, Jackson S T and Nava M-L (2008). Scaling Environmental Change through the Community-Level: A Trait-Based Response-and-Effect Framework for Plants. *Global Change Biology*, **14**, 1125–40.

Thrush S F and Dayton P K (2002). Disturbance to Marine Benthic Habitats by Trawling and Dredging—Implications for Marine Biodiversity. *Annual Review of Ecology and Systematics*, **33**, 449–73.

Thrush S F, Whitlatch R B, Pridmore R D, Hewitt J E, Cummings V J and Maskery M (1996). Scale-Dependent Recolonization: The Role of Sediment Stability in a Dynamic Sandflat Habitat. *Ecology*, **77**, 2472–87.

Tomašových A and Kidwell S M (2017). Nineteenth-Century Collapse of a Benthic Marine Ecosystem on the Open Continental Shelf. *Proceedings of the Royal Society B: Biological Sciences*, **284** (1856), 2017.0328.

Van der Biest K, Meire P, Schellekens T, D'Hondt B, Bonte D, Vanagt T and Ysebaert T (2020). Aligning Biodiversity Conservation and Ecosystem Services in Spatial Planning: Focus on Ecosystem Processes. *Science of the Total Environment*, **712**, 136350.

van der Geest M, van der Heide T, Holmer M and de Wit R (2020). First Field-Based Evidence That the Seagrass-

Lucinid Mutualism Can Mitigate Sulfide Stress in Seagrasses. *Frontiers in Marine Science*, **7**, 11.

Villnäs A, Hewitt J, Snickars M, Westerbom M and Norkko A (2018). Template for Using Biological Trait Groupings When Exploring Large-Scale Variation in Seafloor Multifunctionality. *Ecological Applications*, **28** (1), 78–94.

Villnäs A, Norkko J, Hietanen S, Josefson A B, Lukkari K and Norkko A (2013). The Role of Recurrent Disturbances for Ecosystem Multifunctionality. *Ecology*, **94**, 2275–87.

Villnäs A, Norkko J, Lukkari K, Hewitt J and Norkko A (2012). Consequences of Increasing Hypoxic Disturbance on Benthic Communities and Ecosystem Functioning. *PLoS One*, **7** (10), e44920.

Warwick R M (1980). Population Dynamics and Secondary Production of Benthos. In: Tenore K R and Coull B C, eds. *Marine Benthic Dynamics*, University of South Carolina Press, Columbia, SC, pp. 1–24.

Wenzhofer F and Glud R N (2004). Small-Scale Spatial and Temporal Variability in Coastal Benthic O_2 Dynamics: Effects of Fauna Activity. *Limnology and Oceanography*, **49** (5), 1471–81.

Wildish D and Kristmanson D (1997). *Benthic Suspension Feeders and Flow*, Cambridge University Press, Cambridge, UK.

CHAPTER 10

Biodiversity–ecosystem function

10.1 Introduction

High rates of biodiversity loss (Estes et al., 2011; Harnik et al., 2012; Worm et al., 2006) emphasise the importance of understanding how these changes affect ecosystem function. Biodiversity–ecosystem function (BEF) research grew out of the need to: understand the functional implications of species loss; and highlight how and when biodiversity plays a role in ecosystem function. Thus, BEF relationships link the structural and functional elements of the ecosystems and to offer a way of showing why changes in biodiversity matter. The first challenges in BEF research are deciding how to measure biodiversity (there are many ways, see Chapter 6) and then deciding how to measure function—there are many different functions and forms of measurements (see Chapter 9). This is important not only for our basic understanding of ecology but also because we hope to use these relationships to inform resource management, conservation and ecosystem service assessments (Stachowicz et al., 2007). The BEF research agenda challenges the paradigm that cause and effect relationships are always structured by environmental drivers causing changes in biodiversity. Instead, BEF relationships imply that change in biodiversity can cause changes in the environment and that networks of connections may drive functions as emergent properties (Figure 10.1).

This theoretical focus on feedbacks, networks and emergent properties is ironic as much of the early BEF research was highly reductionist and idealistically mechanistic, making for elegant but sometimes overly simplistic experiments. BEF effects can be additive; individual species contribute to a function whereby more species make for a larger contribution. BEF effects can also be synergistic—where the effect on function is greater than the sum of the effects of individual species. Both are important justifications for conserving biodiversity.

In the 1990s, the BEF research agenda grew out of experiments on how plant species richness (one element of biodiversity) influences plant productivity (one function). A good summary of initial conclusions and research directions by Hooper et al. (2005) highlighted the lack of experiments from marine ecosystems at that time. Despite this lack of early uptake, the types of research and the way soft-sediment ecologists typically sample benthic communities predispose us to address BEF questions. Usually within a core or grab we sample multiple trophic levels and species with multiple functional traits, which are linked to processes associated with biogeochemistry, hydrodynamics or habitat formation (Levin et al., 2001; Snelgrove, 1999; Solan et al., 2004).

Early marine BEF studies were mostly laboratory based, conducted in small-scale mesocosms (such as small buckets) and manipulated a small number of species (usually fewer than five)—see Box 10.1. Apart from defining the form of relationships, BEF effects were assessed on the basis that adding species resulted in a higher functional performance (measured for example by an increased nutrient flux) than the additive effect of the individual species—a phenomenon known as overyielding. Individual experiments often revealed quite different results and often the actual species used in the experiment were much more important than the increase in species per se (see Huston, 1997). Unfortunately, these combinations of species were

Ecology of Coastal Marine Sediments: Form, Function, and Change in the Anthropocene. Simon F. Thrush, Judi E. Hewitt, Conrad A. Pilditch and Alf Norkko, Oxford University Press (2021). © Simon Thrush, Judi Hewitt, Conrad Pilditch, and Alf Norkko.
DOI: 10.1093/oso/9780198804765.003.0010

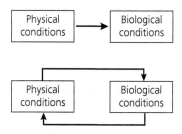

Figure 10.1 The relationship between physical and biological conditions is not always uni-directional, instead BEF relationships suggest feedback loops occur that affect ecosystem dynamics.

Box 10.1 Early BEF studies

Some of the challenges in designing the early BEF experiments were related to the highly reductionist approach to studying biodiversity. Usually a small selection of the available species was selected; these organisms were usually abundant, easy to collect and able to survive under laboratory conditions. Species would then be subjected to the experiment, usually with treatments containing each species in isolation, then combinations of species and then the full assemblage. This tells us what an individual species contributes to the function(s) the experiment is designed to measure, what specific combinations of species contribute and what the collective activity of all the species selected for the experiment contributes. Biomass of organisms is usually kept constant across the experimental treatments to standardise effects.

Factorial permutation of combinations shows how the possible number of combinations quickly rises.

Here is the equation:

$$N!/[R!(N - R)!]$$

where N = number of species and R = group size (combinations of 2, 3, 4 species, etc.).

In a 6-species experiment we have 15 possible combinations of 2 species.

In a 20-species experiment, 190 combinations of 2 species.

In a 60-species experiment, 1770 combinations of 2 species.

As the number of species in our experiment increases, we may also want to increase the number of treatments to investigate the effects of combinations of 3, 4 or more species, further increasing our number of treatments.

Very quickly this type of experiment, which started out being very simple, becomes untenable.

often randomly assembled to restrict any bias in the experimental design. This was unfortunate because disturbance and environmental stress rarely impact on species in a random fashion (Bracken et al., 2008)—more sensitive species or the species that are less resistant or resilient are lost first (see Chapters 3 and 11).

But there were other problems with these early studies, associated with behaviour and the natural history of the experimental species (see Chapter 4). Not all animals behave naturally when stuck in a bucket or a cage. For others the size of the experimental chamber may be wrong, for instance spantangoid urchins, although small, burrow with movement rates of some species exceeding 100 cm d^{-1} (Lohrer et al., 2005). This forces us to realise that if experiments are to inform us about real-world change, we must keep nature at the forefront of our minds. What can we learn from experiments that do not reflect how organisms behave in natural ecosystems?

Another complication with these simple BEF experiments is that many ecosystem functions are not delivered by individual organisms but by patches of organisms and thus the size and density of individuals can contribute to function in a non-additive fashion. Tube mats can stabilise sediments whereas a single tube protruding out of the sediment will enhance erosion (Eckman et al., 1981). The size and fragmentation of patches across the seafloor and their relative location up- or downstream of each other can be important. Habitat structure created by some species affects predation rates on other species (Grabowski et al., 2008). The occurrence of patches dominated by species such as these habitat formers will shape functional relationships both in the patch and potentially in the surrounding area. For example, large sedentary bivalves influence bentho-pelagic coupling in patches and adjacent sediments, whereas their presence in the sediment can limit the movement of mobile species that dominate adjacent sediments (Lohrer et al., 2013). Species that define habitat structure in a patch such as seagrass or shellfish reefs often create a refuge from predation for other species. These species will venture into the surrounding sediments to feed but do not stray far from their refuge, which creates changes in biodiversity and function associated with this zone of fear (Grabowski & Kimbro, 2005; Randall, 1965).

BEF studies had to start somewhere and these 'model system' approaches were tractable, easy to 'rigorously' design and have been informative (Benton et al., 2007). However, this simplification of nature to aid experimentation can blindside you to the applicability of mechanisms at varying scales (Thrush & Lohrer, 2012). Although mechanisms can be tested and resolved in small-scale and simplified systems, as we scale up the results (for use in a management or conservation context) other processes and other forms of interaction can come into play, influencing how biodiversity affects ecosystem function.

10.2 Theoretical considerations and challenges

Similar to research in other ecosystems, early BEF studies associated with marine soft sediments were focussed on defining the form of the relationship between an ecosystem function and species richness (Figure 10.2). These functional relationships provide insight into different elements of ecosystem function. For example: species acting independently of other species in the community imply idiosyncratic relationships: groups of species that contribute to a function can potentially offer some limited resilience leading to stepwise changes or rivet popping; whereas asymptotically increasing curves imply that a few species are doing most of the work in delivering the function. The form of the BEF relationship may change with shifts in community composition, or as space and time scales of sampling integrate over different scales of heterogeneity. We have not studied enough of these relationships to understand if they exhibit consistent patterns or change in predictable ways.

There have been a number of mechanisms proposed to account for positive BEF relationships (Srivastava & Vellend, 2005), including factors that stabilise functional performance. Insurance effects come into play when different species can perform similar functional roles, but these species respond differently to stressors and thus allow net community functionality to remain the same in the face of environmental change. Related to insurance effects

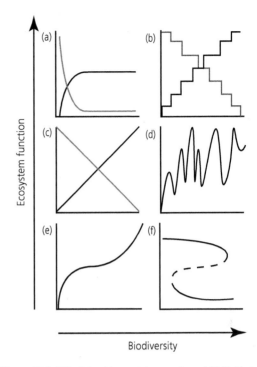

Figure 10.2 BEF relationships can take many forms: (a) initial fast change followed by slower change, (b) stepwise changes, (c) linear change, (d) variable or oscillations around a trend, (e) fast change followed by slower change followed by another increase in rate of change and (f) regime shifts with tipping points and hysteresis.

are portfolio effects; these are generated by the averaging effects of multiple species and do not require any interactions between species. The net effect of the independent fluctuations of many different species is to lower variability, similar to having a diversified investment portfolio that balances variation and risks among different investments. Compensatory dynamic effects occur when species which contribute the same function co-vary out of sync—when one is abundant when the other is not (i.e. negative temporal covariance)—potentially enhancing functional resilience.

As well as these stabilising factors there are factors that can directly contribute to functional performance. Species with more specialist niches might allow diverse communities to be more efficient at exploiting resources, leading to higher functionality (called niche complementarity). Species may help each other out to increase functional performance (called functional facilitation). High-diversity communities are likely to contain many species at low density, limiting their individual effect (called dilution effects). There is also an important effect of the way we sample communities, with the chance of including a dominant, functionally important species increasing with sampling intensity (called the sampling effect). These theoretical concepts highlight that diversity can be important in stabilising functions as well as maximising performance.

One of the values of considering these theoretical concepts early in your research is that it can really help focus your questions and experimental design. Particularly with large-scale and more real-world BEF experiments the results will generally not unequivocally support any one theory, rather they seem to work in combination. It is hard to balance simplicity and complexity, especially when dealing with inherently complex relationships like BEF. Nevertheless, there is evidence that species richness, and richness in genotypes and functional groups are positively related to some ecosystem functions and the strength of these relationships increases both over time and with increasing spatial scale (Stachowicz et al., 2007).

10.3 What elements of biodiversity relate to function?

So far in our discussion of BEF relationships we have focussed on richness, which is the most commonly used measure of biodiversity. As discussed in Chapter 8, species richness is but one of many different measures of biodiversity and we might expect to gain extra insights from using different or multiple measures of biodiversity. The question of what response variable(s) to use in BEF studies is not trivial. We can measure biodiversity or have some index or indicator for biodiversity at any level of biological organisation, from genes to ecosystems. The choice might seem simple if you're interested in genetics or community ecology, but is that the right way to decide? If our research is trying to inform management and conservation then should that not influence our decisions? Gamfeldt et al.'s (2015) meta-analysis of a range of marine experiments supports the earlier conclusions of Stachowicz et al. (2007), but they go on to show that

diversity effects were not as strong as the effect of the highest-performing species. The highest-performing species could be used as an indicator of biodiversity just as we consider iconic species as umbrella species in the conservation of biodiversity. These key species, foundation species, keystone species or ecosystem engineers may well reveal important BEF relationships, especially when we are trying to compare functional performance across habitats or ecosystems. But perhaps the major role of biodiversity is in stability and multifunctionality.

Focussing on the key species makes the mechanistic links much easier to resolve, but without a doubt the real 'biodiversity' aspect of BEF relationships is lost. Removing the key species does not imply that the functional performance will drop to zero; the other species in marine sediments that are left behind are often small surface-dwelling species. This is a zone of intense chemical gradients with a high degree of biogeochemical reactivity, so it would be surprising if these species did not contribute to BEF relationships. All this important context for individual species can get lost in meta-analysis that seeks to find a universal relationship. What do we take from these conclusions? That there is no universal effect, our conclusions are premature, our experimental approach is limited or we should manage seafloor systems like corn farms—homogeneous and single function?

Two other aspects of biodiversity are worth considering: rarity and functional diversity. Rarity is ubiquitous in ecological communities; it is partly an artefact of our size-based sampling designs, but most species are rare (Gray, 2002) (see Chapter 6). Large predators are not sufficiently abundant to be sampled in sampling designs focussed on sampling macrofauna (let alone meiofauna or microbial communities) but they can have profound effects on the functioning of benthic communities. The tail of rare species apparent in SAD and SOD diagrams might also be important in ecosystem function. Species that are rare at one time may change in their abundance or occurrence at another, but there are many potential reasons why individual species might be rare in any particular habitat and community. The insurance effect is often invoked to emphasise the role of maintaining diversity and taking over functional roles when the present functional dominant is impacted. This implies that those species with the potential for functional replacement are more resistant to disturbance or are better suited to coping with environmental or community change (Hewitt et al., 2016). This assumption is linked with the overall number of species exhibiting a functional trait to suggest that a community may have functional redundancy for a specific function (Greenfield et al., 2016; Törnroos & Bonsdorff, 2012). Although we cannot manipulate rarity easily, we can understand its role by looking at changes in community structure and composition over space, time and particularly down environmental gradients. Greenfield et al. (2016) considered both species abundance and occurrence across an extensive area of sandflat and found that specific functional groups of rare species contributed substantially to group diversity, providing insight into resilience.

One of the most useful ways of linking community composition to function is through the use of functional groups. These are groups of species that share important attributes that relate to where and how they live in the sediment, what they eat or what structures they create, how they affect hydrodynamics or how they respond to disturbance and stress. The large diversity of animals living in marine sediments has for a long time had researchers grouping them in terms of their role in the sediment (Rhoads & Young, 1970). Much of the early functional classification focussed on how the animals fed (Fauchauld & Jumars, 1979) or bioturbated (Pearson, 2001). More recently biological traits analysis has been used to identify suites of attributes that can be linked into groups to allow for testing of specific questions (Balvanera et al., 2006), or an overall assessment of functional biodiversity.

Apart from defining trait groups to inform relationships with function, there now are a range of univariate and multivariate functional diversity indices (see Chapter 6). These indices attempt to capture information about different aspects of the multidimensional cloud of points in trait space. Indices capture how much of the trait space is filled with points (functional richness) or describe aspects of the distribution of points within the cloud (such as functional evenness and functional divergence). Tests of different forms of these indices show very variable results (Schleuter et al., 2010), implying the

need for both further methods development and empirical testing against measured functional responses. The use of network analysis of species that share traits is one approach that has shown how the composition and connectivity between clusters of species that share traits change under stress and highlight the multifunctional nature of benthic communities (Siwicka et al., 2020).

10.4 What functions? Singular or multiple?

Soft sediments typically deliver multiple ecosystem functions which means we also need to consider how we factor multifunctionality into our assessment of BEF relationships. Some functions within the sediment are tightly coupled, for example the breakdown of organic matter will affect both nitrogen and carbon processing. Others may be more loosely connected, e.g. sediment stabilisation and sediment production. This multifunctionality is encompassed in soft-sediment experiments where multiple functional response variables are simultaneously measured (e.g. nutrient fluxes, benthic oxygen consumption and primary productivity).

The concept of multifunctionality in BEF experiments has been reviewed by Byrnes et al. (2014). This review highlights how different ways of operationalising multifunctionality can affect interpretation of experimental results and recommends testing whether diversity influences the performance of multiple functions, with the idea that this will lead to the development of functional performance curves. Villnäs et al. (2013) conducted a field experiment involving repeated hypoxic disturbance to the seafloor and measured a range of functions related to nutrient cycling, productivity, ecosystem metabolism and habitat structuring to define functional responses. They determined the effects of increasing disturbance on ecosystem multifunctionality and showed both an earlier and stronger response on overall ecosystem functioning than was evident from any single ecosystem function variable. These findings are important because although a consistent technique to quantify the degree of multifunctionality in communities has yet to evolve, the implication is that as we consider multifunctionality of communities we may expect the functional curves linking measures of diversity to degrees of multifunctionality to be sensitive to environmental change.

The interconnectedness of processes in marine sediments emphasises the need to develop new ways of analysing experimental studies to better capture how these networks are affected by biodiversity, and how they respond to environmental change. As we start to integrate data over broader space and time scales observational studies are likely to be particularly useful in revealing the suite of functions that may be delivered from a particular system, and assessing which ones are likely to dominate depending on the surrounding environmental conditions and context (Hiddink et al., 2009). Gradients in environmental conditions or biodiversity can help you figure out how changes in the gradient factor(s) influence function. Godbold and Solan (2009) took this approach to see how biodiversity changed along a short organic enrichment gradient. The function they were interested in was related to how sediment particles were mixed in the sediment column. They found that species richness was more important (accounting for 34% of the total variability) than total sediment organic carbon (accounting for 5% of the total variability).

10.5 Scaling up BEF relationships

As we have worked through the development of BEF research we have touched on some of the issues associated with relating model system-based experiments to real-world situations, for example in the recognition of the different effects of individual animals vs patches. In this section we want to explicitly focus on the ways we build up our knowledge of BEF relations in seafloor sediments to contribute to conservation and environmental management problems (Thrush & Lohrer, 2012).

Interesting things begin to happen as we start encompassing more and more time or increasing spatial heterogeneity in BEF experiments and often the strength of BEF relationships becomes clearer (Stachowicz et al., 2007, 2008). Essentially the indirect effects that work through interaction networks, and the processes that work across scales (fast and slow, small and broad), become more important. To really address BEF relationships with more than a

very small number of species we need to utilise natural variation in diversity. Working across natural gradients allows us to encompass more time and more spatial heterogeneity. Of course, with this comes the need to measure other co-variables to help account for other sources of variation in our experimental results. Such an approach can help us understand how BEF relationships can vary across the seafloor, which might be important for the stability of functioning at the habitat scale due to processes like complementarity. It also allows us to link to studies of disturbance and question whether hotspots of functional performance change over time or as we stress or disturb particular locations.

One type of study design that follows this approach nests experiments into locations that vary in diversity. Thrush et al. (2017) used this study design to investigate how different measures of biodiversity (species and functional diversity) influenced nitrogen processing in marine sediments. This work began with an extensive survey that was used to identify 28 locations across 300,000 m² of intertidal sandflat that had different combinations of either high or low richness or abundance of a functional group expected to change nutrient processing (through bioturbation, bioirrigation and porewater bioadvection). At each location, experimental plots of three treatments were used to test how the benthic community would respond to different levels of sediment porewater nitrogen concentration. The experiment ran for seven weeks before measures of nutrient and oxygen fluxes, porewater concentrations and macrofaunal community composition were taken. The results did not show a very clear winner in which was the best diversity measure to relate to function; community, functional diversity and the presence of ecosystem engineering species were all important but the relative importance varied with the response variables and the experimental level. However, using the BEF relationships developed from the experiment it was possible to estimate spatial variation in functionality across the whole study site and show how places changed in their contribution with increasing nutrient loading of the sediment (Figure 10.3).

This variability in functionality observed within one community/habitat type has important

surface pore   deep pore   ammonia   DO
water   water   efflux   consumption

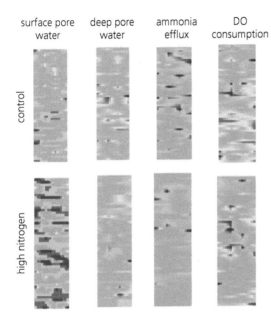

Figure 10.3 Changes in location of functional hotspots (red—high performance) were observed with nitrogen addition. From Figure 3, Thrush et al. (2017). DO: dissolved oxygen.

implications for scaling up functional contributions, realising the limitations of scaling up based on simple scaling coefficients and predicting changes in ecosystem function with environmental change. Models based on theory and mechanistic relationships can be used to scale up BEF relationships, but the empirical research helps to both define relationships and make us more aware of the assumptions we make in building models. Assessing how these mechanisms scale up and underpin function measured at different scales is not a trivial process but essential. A good example is a very simple model that looked at sediment mixing depth and species richness (Solan et al., 2004). Although this model emphasised the importance of individual species that do a lot of work moving sediment rather than species richness per se, it also very successfully highlighted the importance of non-random species loss (due to the link between extinction risk and functional traits). Dealing with these issues of scaling up BEF relationships is a major challenge that requires a substantive and sustained commitment to the research. Demonstrating that biodiversity does matter to ecosystem function will require

combined iterative approaches using both modelling, and experimental and observational studies (Naeem, 2008; Thrush & Lohrer, 2012).

10.6 From BEF to BES

Apart from scaling up BEF relationships in space and time, we can also scale up in terms of moving from a bio-physical assessment into a socioecological assessment. This is the linkage to Biodiversity–Ecosystem Services (BES) (Figure 10.4). Sediments contribute to a wide range of marine ecosystem services, especially in coastal and estuarine ecosystems (Barbier et al., 2011; Barbier, 2012). Services by definition must be valued by people but are ultimately underpinned by ecosystem functions (see Chapter 9). The translation from ecosystem process to functions to services highlights the importance of considering how multiple processes interact. This is a very complex social-ecological problem and we may not have all the information needed to build mechanistic models. In these situations, we may need to rely on expert knowledge to model relationships. Not only are we challenged by defining the relationships between process, function and service but over most of the world we do not have good data on seafloor communities and processes. Thus the use of expert opinion has been used to define ecosystem principles that could be used to map the potential for service delivery (Townsend et al., 2014). Only by nesting ecosystem services into broader social-ecological system interactions can we identify the true value of biodiversity to humanity and identify feedbacks that can lead to change (Snelgrove et al., 2014).

If we hope to advise management and conservation more effectively on the ramifications of biodiversity loss, then we must continue to bridge the chasm between simplified, small-scale experiments and complex, large-scale processes. Here we have focussed on the challenges in developing the links between the fundamental workings of seafloor ecosystems and how they underpin many human needs and values. But there is an equally massive challenge in dealing with the social dimensions of ecosystem services including the monetisation of services and the plurality of human values. Important contributions await in ecosystem-based

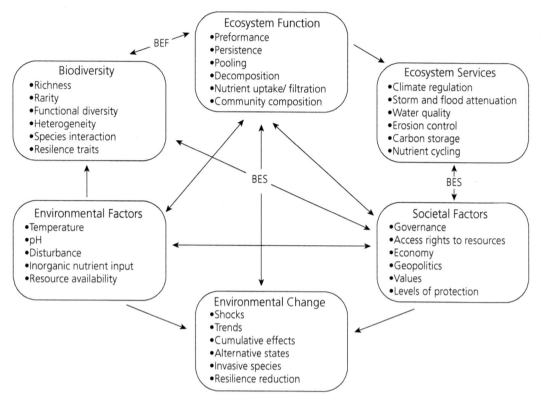

Figure 10.4 Connections between biodiversity, ecosystem function, ecosystem services, society and the environment.

management, resilience-thinking and restoration of degraded ecosystems.

10.7 Close out

There is still a lot of opportunity for research to address how different attributes of biodiversity best relate to ecosystem function (Butterfield & Suding, 2013). Clearly observations of natural systems should inform what measure(s) of biodiversity and what ecosystem function(s) should be considered, just as much as the specifics of theoretical questions. Species richness is easy to calculate and can be easily compared across systems but is a very general measure. Functional traits and diversity are perhaps better linked to addressing more mechanistic questions about specific functions and multiple functions especially in diverse communities, when size or biomass of species can become more important (Duncan et al., 2015). Given that most species in a commu-

nity are rare, we need to develop approaches that allow us to tease apart the effects of specific ecosystem engineer species and overall diversity (Schenone & Thrush, 2020). To enhance relevance to management and society, we need to start to think about diversity at broader landscape scales. In particular, research addressing how the connectivity between habitats or communities relates to driving functions would be a major step forward in assessing the role of cross-scale interactions in affecting seafloor functioning.

References

Balvanera P, Pfisterer A B, Buchmann N, He J-S, Nakashizuka T, Raffaelli D and Schmid B (2006). Quantifying the Evidence for Biodiversity Effects on Ecosystem Functioning and Services. *Ecology Letters*, **9**, 1146–56.

Barbier E B (2012). Progress and Challenges in Valuing Coastal and Marine Ecosystem Services. *Review of Environmental Economics and Policy*, **6** (1), 1–19.

Barbier E B, Hacker S D, Kennedy C, Koch E W, Stier A C and Silliman B R (2011). The Value of Estuarine and Coastal Ecosystem Services. *Ecological Monographs*, **81** (2), 169–93.

Benton T G, Solan M, Travis J M J and Sait S M (2007). Microcosm Experiments Can Inform Global Ecological Problems. *Trends in Ecology and Evolution*, **22** (10), 516–21.

Bracken M E S, Friberg S E, Gonzalez-Dorantes C A and Williams S L (2008). Functional Consequences of Realistic Biodiversity Changes in a Marine Ecosystem. *Proceedings of the National Academy of Sciences of the United States of America*, **105** (3), 924–8.

Butterfield B J and Suding K N (2013). Single-Trait Functional Indices Outperform Multi-Trait Indices in Linking Environmental Gradients and Ecosystem Services in a Complex Landscape. *Journal of Ecology*, **101** (1), 9–17.

Byrnes J E K, Gamfeldt L, Isbell F, Lefcheck J S, Griffin J N, Hector A, Cardinale B J, Hooper D U, Dee L E and Emmett Duffy J (2014). Investigating the Relationship between Biodiversity and Ecosystem Multifunctionality: Challenges and Solutions. *Methods in Ecology and Evolution*, **5** (2), 111–24.

Duncan C, Thompson J R and Pettorelli N (2015). The Quest for a Mechanistic Understanding of Biodiversity–Ecosystem Services Relationships. *Proceedings of the Royal Society B: Biological Sciences*, **282** (1817), 2015.1348.

Eckman J E, Nowell A R M and Jumars P A (1981). Sediment Destabilization by Animal Tubes. *Journal of Marine Research*, **39**, 361–74.

Estes J A, Terborgh J, Brashares J S, Power M E, Berger J, Bond W J, Carpenter S R, Essington T E, Holt R D, Jackson J B C, Marquis R J, Oksanen L, Oksanen T, Paine R T, Pikitch E K, Ripple W J, Sandin S A, Scheffer M, Schoener T W, Shurin J B, Sinclair A R E, Soulé M E, Virtanen R and Wardle D A (2011). Trophic Downgrading of Planet Earth. *Science*, **333** (6040), 301–6.

Fauchauld K and Jumars P A (1979). The Diet of Worms: A Study of Polychaete Feeding Guilds. *Oceanography and Marine Biology Annual Review*, **17**, 193–284.

Gamfeldt L, Lefcheck J S, Byrnes J E K, Cardinale B J, Duffy J E and Griffin J N (2015). Marine Biodiversity and Ecosystem Functioning: What's Known and What's Next? *Oikos*, **124** (3), 252–65.

Godbold J A and Solan M (2009). Relative Importance of Biodiversity and the Abiotic Environment in Mediating an Ecosystem Process. *Marine Ecology Progress Series*, **396**, 273–82.

Grabowski J H, Hughes A R and Kimbro D L (2008). Habitat Complexity Influences Cascading Effects of Multiple Predators. *Ecology*, **89** (12), 3413–22.

Grabowski J H and Kimbro D L (2005). Predator-Avoidance Behavior Extends Trophic Cascades to Refuge Habitats. *Ecology*, **86** (5), 1312–19.

Gray J S (2002). Species Richness of Marine Soft Sediments. *Marine Ecology Progress Series*, **244**, 285–97.

Greenfield B L, Kraan C, Pilditch C A and Thrush S F (2016). Mapping Functional Groups Can Provide Insight into Ecosystem Functioning and Potential Resilience of Intertidal Sandflats. *Marine Ecological Progress Series*, **548**, 1–10.

Harnik P G, Lotze H K, Anderson S C, Finkel Z V, Finnegan S, Lindberg D R, Liow L H, Lockwood R, McClain C R, McGuire J L, O'Dea A, Pandolfi J M, Simpson C and Tittensor D P (2012). Extinctions in Ancient and Modern Seas. *Trends in Ecology and Evolution*, **27** (11), 608–17.

Hewitt J E, Thrush S F and Ellingsen K E (2016). The Role of Time and Species Identities in Spatial Patterns of Species Richness and Conservation. *Conservation Biology*, **30** (5), 1080–8.

Hiddink J G, Davies T W, Perkins M, Machairopoulou M and Neill S P (2009). Context Dependency of Relationships between Biodiversity and Ecosystem Functioning Is Different for Multiple Ecosystem Functions. *Oikos*, **118**, 1892–900.

Hooper D U, Chapin III F S, Ewel J J, Hector A, Inchausti P, Lavorel S, Lawton J H, Lodge D M, Loreau M, Naeem S, Schmid B, Setälä H, Symstad A J, Vandermeer J and Wardle D A (2005). Effects of Biodiversity on Ecosystem Functioning: A Consensus of Current Knowledge. *Ecological Monographs*, **75** (1), 3–35.

Huston M A (1997). Hidden Treatments in Ecological Experiments: Re-Evaluating the Ecosystem Function of Biodiversity. *Oecologia*, **110**, 449–60.

Levin L A, Boesch D F, Covich A, Dahm C, Erséus C, Ewel K C, Kneib R T, Moldenke A, Palmer M A, Snelgrove P A, Strayer D and Weslawski J M (2001). The Function of Marine Critical Transition Zones and the Importance of Sediment Biodiversity. *Ecosystems*, **4**, 430–51.

Lohrer A M, Rodil I F, Townsend M, Chiaroni L D, Hewitt J E and Thrush S F (2013). Biogenic Habitat Transitions Influence Facilitation in a Marine Soft-Sediment Ecosystem. *Ecology*, **94**, 136–45.

Lohrer A M, Thrush S F, Hunt L, Hancock N and Lundquist C (2005). Rapid Reworking of Subtidal Sediments by Burrowing Spatangoid Urchins. *Journal of Experimental Marine Biology and Ecology*, **321** (2), 155–69.

Naeem S (2008). Advancing Realism in Biodiversity Research. *Trends in Ecology & Evolution*, **23**, 414–16.

Pearson T H (2001). Functional Group Ecology in Soft-Sediment Marine Benthos: The Role of Bioturbation. *Oceanography and Marine Biology*, **39**, 233–67.

Randall J E (1965). Grazing Effects on Sea Grasses by Herbivorous Reef Fishes in the West Indies. *Ecology*, **46**, 255–60.

Rhoads D C and Young D K (1970). The Influence of Deposit-Feeding Organisms on Sediment Stability and Community Trophic Structure. *Journal of Marine Research*, **28**, 150–78.

Schenone S and Thrush S F (2020). Unravelling Ecosystem Functioning in Intertidal Soft Sediments: The Role of Density-Driven Interactions. *Scientific Reports*, **10** (1), 11909.

Schleuter D, Daufresne M, Massol F and Argillier C (2010). A User's Guide to Functional Diversity Indices. *Ecological Monographs*, **80** (3), 469–84.

Siwicka E, Thrush S F and Hewitt J E (2020). Linking Changes in Species–Trait Relationships and Ecosystem Function Using a Network Analysis of Traits. *Ecological Applications*, **30** (1), e02010.

Snelgrove P V R (1999). Getting to the Bottom of Marine Biodiversity: Sedimentary Habitats: Ocean Bottoms Are the Most Widespread Habitat on Earth and Support High Biodiversity and Key Ecosystem Services. *Bioscience*, **49** (2), 129–38.

Snelgrove P V R, Thrush S F, Wall D H and Norkko A (2014). Real World Biodiversity–Ecosystem Functioning: A Seafloor Perspective. *Trends in Ecology and Evolution*, **29** (7), 398–405.

Solan M, Cardinale B J, Downing A L, Engelhardt K A M, Ruesink J L and Srivastava D S (2004). Extinction and Ecosystem Function in the Marine Benthos. *Science*, **306**, 1177–80.

Srivastava D S and Vellend M (2005). Biodiversity-Ecosystem Function Research: Is It Relevant to Conservation? *Annual Review of Ecology and Systematics*, **36**, 267–94.

Stachowicz J J, Best R J, Bracken M E S and Graham M H (2008). Complementarity in Marine Biodiversity Manipulations: Reconciling Divergent Evidence from Field and Mesocosm Experiments. *Proceedings of the National Academy of Sciences of the United States of America*, **105**, 18842–7.

Stachowicz J J, Bruno J F and Duffy J E (2007). Understanding the Effects of Marine Biodiversity on Communities and Ecosystems. *Annual Review of Ecology and Systematics*, **38**, 739–66.

Thrush S F, Hewitt J E, Kraan C, Lohrer A M, Pilditch C A and Douglas E (2017). Changes in the Location of Biodiversity–Ecosystem Function Hot Spots across the Seafloor Landscape with Increasing Sediment Nutrient Loading. *Proceedings of the Royal Society B: Biological Sciences*, **284** (1852), 2016.2861.

Thrush S F and Lohrer A M (2012). Why Bother Going Outside: The Role of Observational Studies in Understanding Biodiversity–Ecosystem Function Relationships. In: Paterson D M, Solan M and Aspenal R, eds. *Marine Biodiversity Futures and Ecosystem Functioning: Frameworks, Methodologies and Integration*, Oxford University Press, Oxford, UK, pp. 198–212.

Törnroos A M and Bonsdorff E (2012). Developing the Multitrait Concept for Functional Diversity: Lessons from a System Rich in Functions but Poor in Species. *Ecological Applications*, **22**, 2221–36.

Townsend M, Thrush S F, Lohrer A M, Hewitt J E, Lundquist C J, Carbines M and Felsing M (2014). Overcoming the Challenges of Data Scarcity in Mapping Marine Ecosystem Service Potential. *Ecosystem Services*, **8**, 44–55.

Villnäs A, Norkko J, Hietanen S, Josefson A B, Lukkari K and Norkko A (2013). The Role of Recurrent Disturbances for Ecosystem Multifunctionality. *Ecology*, **94** (10), 2275–87.

Worm B, Barbier E B, Beaumont N, Duffy E J, Folke C, Halpern B S, Jackson J B C, Lotze H K, Micheli F, Palumbi S R, Sala E, Selkoe K A, Stachowicz J J and Watson R (2006). Impacts of Biodiversity Loss on Ocean Ecosystem Services. *Science*, **314**, 787–90.

PART V

Anthropocene

CHAPTER 11

Human impacts

11.1 Introduction

With about half of the world's population living within 100 km of the coast, and an increasing number of people relying on the oceans for their livelihood, human impacts on marine ecosystems are increasing at an accelerating rate. The world's human population continues to grow and is projected to reach 8.6 billion in 2030, and to increase further to 9.8 billion in 2050 and 11.2 billion by 2100 (United Nations, 2017). It is not just the number of people that matters; it is our level of consumption. The ecological footprint of humanity indicates we are now using about two planets' worth of natural resources.

The human dominance of our biosphere has been recognised as a new geological era in earth's history termed the Anthropocene, because not only is our level of consumption increasing, but so are the types of uses we make of the environment. Human uses range from extraction of resources (e.g. harvesting single species, mining minerals, oils and gases) and use of space (shipping lanes, aquaculture, tourism) to disposal of waste products (intended and unintended) and occur from local to global scales. Often these uses have unintended direct consequences such as overfishing or eutrophication and, even more frequently, unintended indirect consequences, such as invasions of exotic species, loss of essential habitats and degradation of the ecosystem services. In the coastal zone, impacts not only occur from users of the marine system, but also from users of the land and freshwater. For example, the modification and destruction of watersheds result in floods and burial of natural coastal

under terrestrial and organic-rich sediments, which in tandem with increased nutrient loading result in eutrophication and spreading deoxygenation (Jackson, 2001; Thrush et al., 2004). Not all uses are negative; eco-tourism and recreation allow people the chance to appreciate, and interact with, nature and drive many of the recent efforts at restoration (see Chapter 13).

Anthropogenic disturbances and stressors have already caused broad-scale changes to soft sediments worldwide by modifying or even eliminating natural habitats and influencing the health, abundance and distribution of functionally important species. These impacts are intense in the coastal zone, which are some of the most valued by humans (Costanza et al., 1997). The list of emerging contaminants continues to grow, as does our understanding of indirect effects on the ecosystem. Invasive species, noise and changes to dispersal patterns join the list of potential stressors. The results are that it is impossible to consider, in a single chapter, all the impacts humans have on the coastal environment. Climate change is of course a major anthropogenic change—we touch on that in Chapter 12. This chapter therefore focusses on some of the key concepts for considering how humans impact the environment and how we might study these impacts. Specifically, we will discuss direct and indirect effects, impacts at global scales and multiple stressors and cumulative effects. In these sections we will highlight the potential for unintended or unrecognised consequences as illustrated by considering specific stressors in specific locations. Finally, we will consider how we might study and manage impacts in these multi-use but vulnerable systems.

Ecology of Coastal Marine Sediments: Form, Function, and Change in the Anthropocene. Simon F. Thrush, Judi E. Hewitt, Conrad A. Pilditch and Alf Norkko,
Oxford University Press (2021). © Simon Thrush, Judi Hewitt, Conrad Pilditch, and Alf Norkko.
DOI: 10.1093/oso/9780198804765.003.0011

11.2 Direct and indirect effects; unintended consequences

With a vast array of different stressors, we need to look for generality in the way they impact seafloor ecology. Stressors vary in their direct effects, for example:

- physical disturbance
- species removal
- alteration/removal of habitat
- species addition—generally these are invasive species
- biodiversity loss
- contamination which includes behavioural changes and toxicity
- changes to movement/connectivity
- alteration of food quantity and quality
- changes in chemical balances and elemental cycles.

Looking at this type of categorisation allows an easy way to incorporate single human activities that produce more than one stressor (Figure 11.1). It also provides a way to incorporate single stressors that produce more than one direct effect. For

example, small particles of plastics (microplastics) can reduce food quality, whereas larger particles can create behavioural changes and death (e.g. by strangulation).

Compared to natural disturbances, humans often change the disturbance regime to increase the extent, frequency and magnitude of disturbance (Chapter 3). But human activities can also focus on particular habitats or species, for example, exploitation of particular biogenic habitats comprised of specific functional groups (e.g. suspension feeders) or particular trophic levels (e.g. predatory crustaceans or fish), resulting in their functional elimination.

This, in turn, can result in feedbacks that modify the impact of natural disturbance regimes. For example, the loss of biogenic habitats that stabilise the seafloor can change sediment resuspension dynamics, amplifying the impact of storm events, with subsequent effects on benthic–pelagic coupling and key ecosystem functions (e.g. carbon and nutrient transformation and retention). Tomimatsu et al. (2013) suggested that stressors (especially those derived from human activities) affect ecosystem functions through abiotic effects on the environment

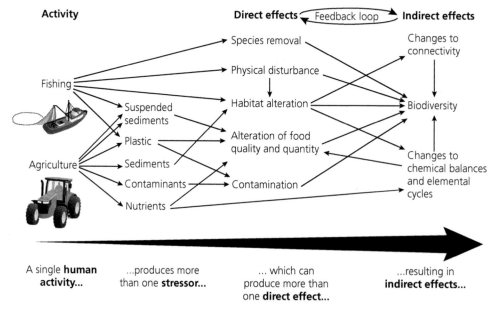

Figure 11.1 A single human activity can produce multiple stressors and affect the ecosystem in multiple ways. With tractor and fishing boat illustrations by Tracey Saxby, IAN Image Library (https://ian.umces.edu/imagelibrary/).

and biotic effects on the ecosystem state, i.e. any biotic variables potentially affecting ecosystem function. These three components (abiotic environment, community and ecosystem state and ecosystem function) can influence one another, and through indirect effects create positive or negative feedbacks that affect ecosystem dynamics.

Increasingly, we are also understanding that identifying effects of stressors, especially contaminants and eutrophication, on ecosystem function and status requires consideration of a full range of ecosystem components from microbes to megafauna. To date microbial responses have been less studied than other aspects (Johnston & Roberts, 2009), although stress-affected microorganisms can shift energy resources to survival rather than growth, significantly altering energy and nutrient flows in the ecosystem (Schimel et al., 2007).

11.2.1 Direct and indirect effects on biodiversity

Both direct and indirect effects are implicated in biodiversity loss. Major problems in deriving generalities include: different species responding at different stress levels (Ugland & Gray, 1982), altering both the number of species and how abundances are distributed across species; and different responses between alpha, beta and gamma diversity (see Chapter 6). For example, Hewitt et al. (2010) observed first an increase and then a sharp decrease in average species richness in response to metal contamination. However, the effect on the heterogeneity of species (beta diversity) at a site was a strong and immediate decrease, while the effect on total species richness at the site, although less strong, was also an immediate decrease.

Indirect effects of human activities on biodiversity, through alterations to the physical and chemical environment and removal of habitat, are common. Loss of key species can generate cascading biodiversity effects down through the organisational classes of species living in soft sediments (macroalgae, epifauna, macro infauna, meiofauna, microphytobenthos and bacteria) or across the functional trait biodiversity, impacting on functional redundancy and resilience (Box 11.1). In some cases,

---

Box 11.1 The positive aspects of invasive species—when bad is good!

The results of indirect effects and feedbacks, although unintended, are not necessarily detrimental.

Marenzelleria, a deep-burrowing, invasive spionid polychaete, has spread rapidly and is now dominant throughout coastal regions of the Baltic Sea, especially in areas prone to hypoxia (Maximov, 2011; Kauppi et al., 2015). *Marenzelleria* has been suggested to potentially counteract hypoxia formation in the bottom water (Conley et al., 2007; Josefson et al., 2012), through a complex interaction of direct and indirect effects. Transport-reaction modelling based on field results suggests that *Marenzelleria*, through bioturbation and irrigation, oxygenates the deeper sediments and therefore has the potential to enhance phosphate retention in sediments. This may alleviate phosphate release from bottom waters and counteract surface-water eutrophication and hypoxia (Norkko et al., 2012). The model suggests that over time bioirrigation leads to a substantial increase in the iron-bound phosphate content of sediments, while reducing the concentration of labile organic carbon. Importantly, the modelling results suggest that the positive feedback mechanism is density dependent (Norkko et al., 2012). Thus, although non-native invasive species are often viewed as stressors that degrade biodiversity, in species-poor areas, or where a different stressor has removed species that provide an important function, invasive species can be a positive influence.

---

sufficient loss of redundancy in trait groupings can lead to functional 'extinction'. This is particularly likely to occur through the loss of larger organisms.

11.3 Local to global impacts; unrecognised consequences

Human activities vary in spatial and temporal scale. Some activities are very discrete, such as the point source of a discharge from a single industry, but may occur more or less continuously. Others are more diffuse, occurring over large areas; these may be event driven (e.g. increased terrestrial sediment inputs into the coastal zone) or, again, more continuous (e.g. nutrient enrichment). Increasingly these days we have to contend with global impacts such as climate change (see Chapter 12), over-fishing, hypoxia and, microplastics. The impacts recognised as being globally important have changed markedly over time, both as activities have changed and as the global consequences have become clearer. The next four examples explore what is presently known about the effects of four globally acknowledged human actions, and the extent of those effects.

11.3.1 Organic enrichment and hypoxia

Eutrophication generally enhances sediment biomass production at early stages of nutrient loading and organic enrichment (i.e. increased food availability), followed by community impoverishment or complete loss when severe hypoxia and anoxia develop (Pearson and Rosenberg, 1978; see Chapter 3). The Pearson and Rosenberg model was developed using information from broad-scale enrichment of muddy subtidal sediments and provides a useful context for interpreting organic enrichment effects and responses to hypoxia. Increasing organic enrichment eliminates bioturbating macrofauna and the species capable of assimilating and transforming the organic matter, resulting in negative feedbacks that intensify the formation of hypoxia and anoxia in the sediments and the water column (i.e. surplus organic matter is burned at the sediment surface rather than assimilated). Depending on the intensity of organic enrichment, increasing amounts of resources are made available to potential colonists following disturbance. The magnitude of population increases by opportunistic species has been shown to depend on a legacy of disturbance and has important implications for transformation and retention of organic matter. Increasing dominance by short-lived opportunistic species at the expense of large and long-lived fauna fundamentally changes the stability of soft-sediment ecosystems, resulting in an ecosystem state characterised by rapid turnover of carbon and nutrients, and a dominance of microbial processes (see Chapter 1).

It is well established that severe hypoxia and anoxia change the behaviour and physiology of, and ultimately kill, benthic fauna resulting in

simplified foodwebs and the dominance of microbial processes that characterise 'dead zones' (Buck et al., 2012; Diaz & Rosenberg, 2008). In 2008, a review by Diaz and Rosenberg (2008) reported dead zones in the Baltic, Kattegat, Black Sea, Gulf of Mexico and East China Sea. Recently Altieri and Diaz (2019) reported exponential increases in the number and severity of dead zones across the globe and the prediction is that these will continue to increase with warming of our oceans (Breitburg et al., 2018).

It has been suggested that eutrophication has plagued the Baltic Sea, which now boasts the largest 'dead zone' in the world, since at least the medieval warming period (950 to c.1250 AD). The first anthropogenically induced hypoxic events are suggested to have taken place following a rapid population increase in Scandinavia, concurrent with clearing of the catchment and the development of more efficient farming practices (Zillén and Conley, 2010). It was not, however, until the 1960s that large-scale hypoxia developed following a major population increase in the catchment, combined with the introduction of intensive agriculture practices including artificial fertilisers (Figure 11.2, Carstensen et al., 2014). Today the Baltic Sea has a

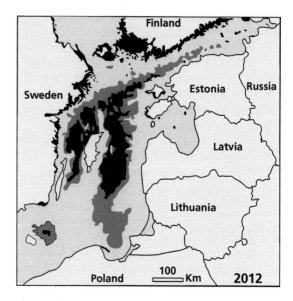

Figure 11.2 Spatial extent of bottom oxygen concentrations < 2 mg· ·L^{-1} (red) and concentrations < 0 mg · L^{-g} (black) in the Baltic Sea in 2012 (adapted from Figure 2, Carstensen et al., 2014).

hypoxic (<2 mg l^{-1}) area that covers up to 70,000 km^2 of the open Baltic Sea with increasing occurrences of seasonal hypoxia also being observed in coastal areas (Conley et al., 2011).

11.3.2 Over-fishing and bottom disturbance

Estimates of the global extent of bottom fishing are hotly disputed. However, there is a long history of studies on the consequences of fishing, beyond those on the effects of over-fished species, changes to the size and structure of marine foodwebs, and impacts of by-catch. These broader consequences include shifts in community and habitat structure, and loss of ecosystem function, particularly as a result of fisheries that disturb the seafloor, such as bottom trawling and dredging (Figure 11.3).

Moreover, the constant disturbance of the seafloor can also result in changes in sediment types. Oberle et al. (2016) demonstrated large-scale change in sediment type, with increased fine sediment, whereas Puig et al. (2012) observed removal of bedforms. The mass resuspended by bottom trawling globally over the continental shelfs is calculated to be roughly equivalent to the mass of riverine-input sediment (Oberle et al., 2016).

For many years, studies struggled with the mismatch between the results from small-scale and short-term experimental manipulations assessing fishing impacts and the broad scale of the fisheries' activity (Thrush et al., 2015). Effects found were often habitat- and trawl gear-specific (Collie et al., 2000; van Denderen et al., 2014). Despite these problems, accumulated evidence from numerous studies has highlighted consistent patterns: the loss of large and long-lived organisms, decreases in habitat heterogeneity (see section 11.4) and species diversity, and the loss of important functional groups (e.g. suspension feeders). For example, bottom fishing removes large and emergent benthic organisms capable of enhancing habitat heterogeneity over large areas of the seafloor (Dayton et al., 1995; Mangano et al., 2014; McConnaughey et al., 2000). In 2012, de Juan and Demestre (2012) developed an indicator of bottom fishing on benthic ecosystems that focussed on these losses. These non-random losses of species can result in a loss of ecosystem function and services.

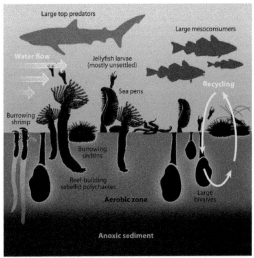

(a) Former state

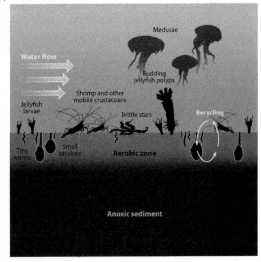

(b) More fished world

Figure 11.3 Ecosystem consequences of extensive fishing (Figure 3, Thrush and Dayton, 2010). Reproduced with permission from the *Annual Review of Marine Science*, Volume 2 © 2010 by Annual Reviews, http://www.annualreviews.org.

11.3.3 Terrestrial sediment inputs

Prior to the 1990s, the potential for terrestrial sediment inputs to adversely affect coastal soft sediments was largely unrecognised (but see McKnight, 1969 and Peterson, 1985). Instead excessive terrestrial sedimentation was thought to occur in limited locations and to largely affect rocky reefs. Fauna of soft sediments were expected to be robust, especially when the sediment type did not change. However, increased delivery of sediment to coastal areas was highlighted by GESAMP (1994). Not long after, a study in the highlands of Sri Lanka reported that the conversion of forest to agricultural land had increased the rate of sediment run-off by two orders of magnitude, up to 7,000 metric tons per km² per year (Hewawasam et al., 2003). Terrestrial sediment inputs are increasingly recorded as entering the coastal environment as the frequency of storms and floods increases with changing climate patterns.

Long-term and large-scale implications for biodiversity and ecosystem function through sediment deposition, increased suspended sediment, decreased water clarity and changes in the extent and fragmentation of habitats are apparent (see Figure 11.4). Experimental studies have demonstrated direct and indirect effects of smothering and suspended sediment on faunal community composition, microphytobenthos and seagrass, through

changes to mobility, food quality, light availability and oxygenation of sediments. Ongoing changes from sandy to muddy sediments alter colonisation and recruitment patterns, fragment sandy sediments and increase turbidity, impacting macroalgae, seagrasses and suspension feeders including shellfish and sponges, all of which are habitat-structuring organisms, over large areas of the seafloor (Karlson et al., 2002; Rabalais et al., 2002; Thrush et al., 2004). Sandy habitats are finally replaced with large extensive mudflats with high rates of sedimentation.

11.3.4 Plastic pollution

Even more recent is the recognition of the potential effects on the environment of our love affair with plastic. Although plastics have conferred multiple benefits on society from improvements in human health to reductions in transport costs, the environment suffers from their slow degradation and volume. With global plastic production over 300

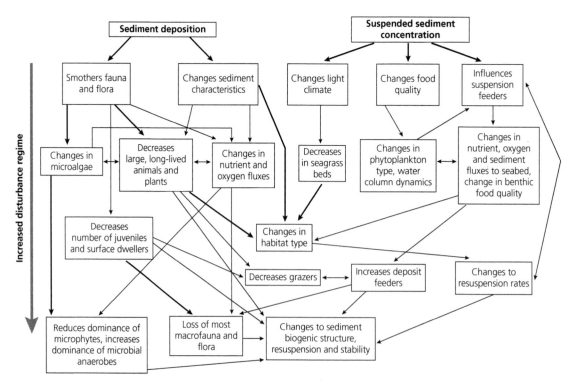

Figure 11.4 Direct and indirect effects of terrestrial sediment inputs into estuaries (Figure 5, Thrush et al., 2004).

million metric tons per year (Law, 2017), plastic in various forms and sizes has been detected around the world, from the Arctic to the Antarctic in all major marine habitats (Law, 2017). Unfortunately, few reviews of the production, fate and impacts of plastic (e.g. Law, 2017; Wang et al., 2016; Cole et al., 2011) focus on the marine benthic environment. Early marine work focussed on the pelagic environment, or plastic's encounters with fish and birds, as much of the plastic debris is light and floats. Today, large numbers of microplastics have been recorded in beach sediments, with the potential to change sediment grain size and thus permeability, diffusivity and nutrient fluxes (Carson et al., 2011; Cluzard et al., 2015). Ingestion of particles has been recorded for a large range of benthic macrofauna (see references in Wang et al., 2016); however, there have been few demonstrations of ingestion leading to population or community changes (but see Green et al., 2017). Death of individuals and subsequent changes to populations of communities are also likely to result from toxic responses to microplastic-associated contaminants and other pollutants that are attracted to the microplastics (Cole et al., 2011).

11.4 Multiple stressors and cumulative effects

As with natural disturbances (see Chapter 3), the stressors and disturbances generated by human activities have a strong spatio-temporal dimension, meaning that they may not only act across a range of spatial and temporal scales, creating legacies in the landscape, but also they often are connected in time and space. This, together with high use of systems, means that it is rare for a place, species or community to be affected by a single activity or stressor. Instead multiple stressors accumulate over a range of space and time scales resulting in cumulative effects.

Cumulative effects over time of a single stressor depend on the intensity of the stress, the time period available for adaptation and on whether indirect effects and feedbacks result in non-linear responses (see section 8.5). Organisms that are already under stress may be unable to cope with more stress (Lenihan et al., 1999) or be adapted to stress (Maltby, 1999); both of these can create

thresholds in response. But if predicting the impact of more of the same stressor is difficult, predicting cumulative effects of multiple stressors is much harder. In a simple world the effects of the two stressors could be added together (e.g. Halpern et al., 2007). However, studies on systems around the world demonstrate that multiplicative interactions are common (Ban et al., 2014; Thrush et al., 2008). These interactions may be either synergistic (larger than the sum of the individual effects) or antagonistic (where the effect of one stressor dampens the effect of another). Thrush et al. (2008) observed that multiplicative effects on abundances of infauna were generally antagonistic between sediment characteristics (mud and coarse sediment) and synergistic between mud and contaminants. Increasingly, synergistic effects are being reported, although this is highly dependent on the stressor combination (Crain et al., 2008).

Jackson et al. (2001) provide an account of how humans have had a profound impact on marine resources over millennia. Exploring the history of change in a number of contrasting coastal ecosystems, they illustrate how a sequence of cumulative human impacts alter marine ecosystems. They argue that fishing (targeting predators and key functional groups) often preceded all other human impacts, increasing the vulnerability to stressors such as eutrophication, habitat destruction, spread of invasive species and changes in climate regimes in a chain of events (Figure 11.5).

Below we give two examples of the cumulative effects of multiple stressors: the global problem of habitat fragmentation and homogenisation; and accumulating impacts over time changing the ecosystem of the Chesapeake Bay.

11.4.1 Habitat fragmentation and homogenisation

There is increasing evidence that seafloor habitats are being firstly fragmented and finally homogenised throughout the world's oceans (Figure 11.6, although this can also be seen in Figure 11.3). This results from numerous activities generating cumulative and multiple stressors over a range of scales. For example, as we discussed above, enrichment and the resulting hypoxia reduces the types of

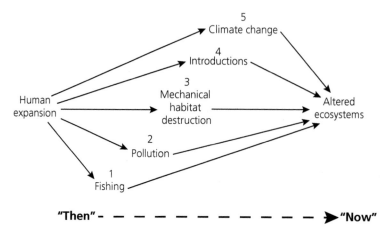

Figure 11.5 The sequence of effects is likely to prove important for the impacts subsequent stressors have on a system (Figure 3, Jackson et al. (2001). Reprinted with permission from AAAS).

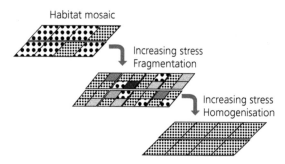

Figure 11.6 Increasing stress leads initially to fragmentation of some habitats and finally to homogenisation when only a single habitat remains.

species that can inhabit large areas of the seafloor, bottom disturbance by fishing removes benthic habitat formers and changes sediment types towards more homogeneous areas of fine sediment, and fine terrestrial sediment deposition smothers or chokes many epifauna, reduces habitat for many epiflora and increases the areas of fine sediment.

Fragmentation of biodiverse habitats creates barriers to maintenance of populations, especially for those species requiring extensive and high-density beds of individuals to provide recruits (Chapter 3). Fragmentation also affects the resilience of habitats by affecting the potential for recovery from small-scale events (Whitlatch et al., 2001). Initially the relation between diversity and habitat heterogeneity results in an increase in diversity with fragmentation at some scales. However, although habitat heterogeneity increases diversity, any specific habitat often needs to cover a certain amount of area for it to effectively offer a habitat to specialist species. This creates a unimodal relationship between fragmentation of specific habitats and communities and their biodiversity, with decreases occurring well before the habitat disappears. de Juan and Hewitt (2011) observed decreases in all aspects of biodiversity with increasing fragmentation of both vegetated and non-vegetated intertidal habitats (decreasing patch size or increased number of patches).

Box 11.2 The Chesapeake as an example of historical change and cumulative impact

The temporal sequence of human activities is often central for understanding the current environmental/ecological status of marine ecosystems as direct and indirect far-field effects are common. For example, changes in catchments that result in increasing sediment and nutrient delivery to coastal seas, concomitant with offshore exploitation of species at high trophic levels (including apex predators), can have synergistic effects on coastal soft-sediment communities and habitats through modification of top-down and bottom-up processes.

With a watershed spanning ~166,000 km², the Chesapeake Bay is the largest estuary in the continental United States and once supported a massive commercial oyster fishery. Human impacts started early, however, with overfishing of top predators in the sixteenth and seventeenth centuries. Coincidently widespread clearing of the catchment for settlement and agriculture increased sedimentation and burial of carbon in the Chesapeake estuary sediments in the mid-eighteenth century. As a result, shifts in primary producer community composition, an early sign of eutrophication, appeared in the late eighteenth century (Cooper & Brush, 1993). Declines in seagrasses and benthic diatoms and concomitant increases in pelagic phytoplankton shifted the base of the foodweb (Officer et al., 1984).

Massive beds and reefs of suspension feeders, primarily oysters, could control a large proportion of the increased turbidity and phytoplankton production until mechanical harvesting of oysters reduced their population by the 1920s (Rothschild et al., 1994). Harvests decreased by as much as 50% by the early 1900s and 98% by the early 1990s (Rick et al., 2016). Following the collapse of the oysters, phytoplankton populations increased dramatically, enhancing the flux of organic matter to the seafloor, resulting in losses of benthic fauna. In the 1930s hypoxia and anoxia started spreading. Remaining oysters were struck by oyster parasite outbreaks in the 1950s, possibly because of exposure to multiple stressors, including sedimentation and hypoxia. Current populations are thought to be less than 1% of their historical size (Rick et al., 2016).

The major loss of suspension feeders serves as an example of how the functional removal of species capable of exerting top-down control on the ecosystem may result in a regime shift in marine soft-sediment systems, similar to those described in lakes (Carpenter, 2003). Hence, cumulative interacting disturbances can make recovery from a degraded ecosystem state difficult. Expensive attempts to restore native oyster populations have not been successful due to the interacting effects of eutrophication, disease, hypoxia and continued dredging that prevent the recovery of oysters and associated communities. However, more recent reports of trend-reversals in eutrophication status and the return of, e.g. submerged aquatic vegetation (e.g. Lefcheck et al., 2017), together with the success of artificial reefs (Lipcius & Burke, 2018), could possibly also facilitate efforts at restoration of suspension feeders in the future (see also Chapter 13).

Many coastal seas are affected by similar chains of events, where a history of early fishing down the foodweb, combined with catchment development and intensification of agriculture, has degraded soft-sediment communities. Because of their inherent characteristics, soft sediments are often places where organic matter accumulates and hence they are at the receiving end of the multiple stressors and their cumulative impacts. Intensifying climate change, with the potential for elevated inputs of sediments and nutrients to coastal seas, has the potential to act synergistically with increasing temperatures to erode the biodiversity of soft sediments.

Homogenisation occurs once the small fragments of previous habitats disappear; species become functionally extinct; biodiversity decreases as only those species able to live in the single extensive habitat remain; and ecosystem services are lost.

11.5 How can we study impacts in the world of cumulative effects and sliding baselines?

11.5.1 Sliding baselines

The concept of sliding baselines was first proposed in the 1990s (for fish communities by Pauly, 1995; and for marine benthos by Dayton et al., 1998), suggesting that our ability to determine what is 'natural' is limited. Dayton et al. (1998) observed that it

was impossible to define a meaningful benchmark against which to assess changes in a kelp forest because many of the large animals had been gone for possibly decades and the presence of previous regime shifts was unknown. Analyses of middens, diaries, letters, photographs and traditional knowledge of indigenous people reveal large differences between the marine world as we see it today (or even in the past 50 years) and as it was in past times (Dayton et al., 1998). This is particularly the case for estuaries and coastal areas with a long history of civilisation (Lotze et al., 2006) (Box 11.2). Beck et al. (2011) document the decline of the oyster reefs which once dominated many estuaries, and Robinson and Frid (2008) discuss the effect of hundreds of years of fishing on the functionality of the North Sea. Unfortunately, the rate of these changes has dramatically accelerated over the past 150–300 years

(Lotze et al., 2006), changing our systems such that our concepts of natural variability and ecosystem functioning may be well wide of the mark.

There are two other important implications of 'sliding baselines'. One is that we should not think of these changes as necessarily being a linear 'slide'. In reality, the slide is likely to have comprised times when no changes appeared to be happening, followed by very sharp changes—tipping points are more likely the norm than not. Secondly, many of our predictive techniques and management activities rely on setting limits based on linear change and an understanding of natural variability. If tipping points are prevalent, not only is the association of the 'dose' and the 'response' incorrect and not only are reference sites possibly incorrectly chosen, but what we think of as 'natural variability' may instead be the enhanced variability expected prior to a tipping point occurring (see Chapter 8).

11.5.2 Integrating multiple studies

It is fairly clear that we cannot create manipulative studies that encompass multiple stressors at the time and space scales over which they create consequences. We also cannot study all ecosystem components and their connections. But what we can do is build information from a series of studies (as first proposed by Eberhardt & Thomas, 1991) and clever use of models.

de Juan et al. (2018) propose an approach to the measurement of Ecological Integrity based on multiple studies and lines of evidence. They illustrate the framework's effectiveness in three case studies, including a data-poor situation, a situation where there is a lack of reference sites, and an area with diffuse sources of stress. For example, in the area with a lack of reference sites, a series of scientific surveys collected data on benthic communities from fishing grounds in the NW Mediterranean, across different levels of trawling effort. Analysis was conducted to test specific hypotheses related to the structure and function of soft-bottom marine ecosystems (de Juan et al., 2007), allowing identification of effects (de Juan & Demestre, 2012). Changes in benthic habitats were observed: homogenised muddy sediment and increased turbidity were found in heavily trawled sites (Palanques

et al., 2001). A further survey based on differences in sediment composition and trawling disturbance levels compared responses to trawling within and across habitat types, allowing robust identification of effects (de Juan et al., 2013). An 'easily interpreted' indicator of the ecosystem status in trawling grounds (e.g. de Juan & Demestre, 2012) was developed. Finally, the observed responses were related to alterations in the provision of services by benthic communities, particularly nursery habitat provision for commercial species (Muntadas et al., 2014, 2015).

11.5.3 Non-random species loss

In soft-sediment systems we know that many species create habitats for other species and that species at risk are often those with specific biological traits. Thus, as many human activities result in the removal of specific habitats and the expansion of others, biodiversity loss is most likely not random. This poses the risk of underestimating biodiversity loss based on species–area relationships as these curves are generally derived by assuming that species richness in the area of interest does not exhibit any spatial pattern (Seabloom et al., 2002), including those generated by species–species or species–habitat interactions. In a first attempt to create a method for assessing non-random species loss, Thrush et al. (2006) utilised a species-habitat accumulation model to predict not only loss of species richness with decreasing number of habitats, but also the effects of losing specific habitats, such as those at risk from bottom fishing. They observed that removing habitats defined by large emergent sedentary species resulted in higher levels of loss than random removal of habitats. They also observed that, in the two locations they sampled, numbers of habitat specialists were not distributed randomly across habitats and that habitat removal demonstrated a consistent loss of large individuals, epifauna, surface dwellers and deposit feeders. The relative proportion of habitat specialists vs habitat-generalists drove variation in both the rate of species accumulation associated with the proportion of species common across all habitats and the impact of habitat homogenisation on species richness. Moreover these two processes interacted, such that the fewer habitat-generalists there were, and the more habitat-specialists, the greater was the prediction of biodiversity loss with decreasing numbers of habitats.

11.5.4 Biological traits

The examples given in the last two sections all highlight the utility of knowledge of biological traits. Species' biological traits (e.g. suspension feeding, being highly mobile) control their responses to their surrounding environment, their interactions with other species, their ability to affect their environment and their ability to recover from disturbances. Biological trait analysis (BTA) can be used to assess whether an impact occurs and to predict the likely flow-on effect to ecosystem functioning (Oug et al., 2012) (see Chapters 9 and 10).

Conversely, traits can also be used to assess the sensitivity of habitats or areas to a planned activity (Clark et al., 2016; Zacharias and Gregr, 2005) when only the traits that control the ability of an organism to withstand a stressor are focussed on (e.g. Tyler-Walters et al., 2009). In order to use BTA to assess sensitivity, general characteristics that predict both the sensitivity to and the resistance to a stressor need to be known. Because of the large number of species present in most marine environments, generalities from the theory of disturbance/recovery are more likely to be used than specific manipulative studies or impact assessments. For example, the Pearson and Rosenberg characterisation for different stages of organic enrichment in marine systems based on sediment dwelling depth, burrowing activity, body size, mobility and life span (Pearson & Rosenberg, 1978). Similarly, Thrush and Dayton (2010) summarised the biological traits that would be affected by bottom fishing. Hewitt et al. (2018) developed a decision tree methodology for predicting sensitivities to the general activity categories of extraction, sedimentation and suspended sediment based on BTA. This method was effective in separating sensitivities to the different stressors while not claiming that all communities differed in sensitivity to the same stressor.

However, there is another part to assessing impacts, whether to a single species or ecosystem component or to ecosystem functions and services,

and that is recovery. If impacts can be recovered from, how quickly can this occur? Traits at the individual scale often define the initial response, but attributes at the population and landscape scales influence recovery, for example, when, how far and how often species are likely to disperse. Species that are highly mobile within the water column on a daily basis such as some benthic fish species, crabs and corophids have a greater potential to colonise a disturbed area than a sedentary attached epifaunal species that reproduces for a limited period infrequently. However, the influence of these traits occurs in combination with location-specific spatial and temporal species rarity and occupancy in the landscape, and population and landscape connectivity. A recent paper by Gladstone-Gallagher (2019) uses traits to consider the relative roles of intensity and spatial and temporal extent of disturbances (stressor footprint) on ecological resilience. They find that the roles of these various factors combined with the location-specific community type drive not only resistance and recovery from disturbance but, as a result, the ecological response footprint (intensity and spatial and temporal scales). These three components are often the components that separate natural from human-driven stress.

11.6 Managing impacts in a multi-use world

Management in a multi-use system, where indirect effects are common and the responses to multiple and accumulating stressors are non-additive, is not easy. In fact, it is easier to create a list of 'don'ts' than to give constructive advice. For example, don't manage on a sectorial basis, don't overlook nonlinear interactions. Thrush et al. (2016) not only highlight the challenges we face but also provide advice, beginning with the simple, 'Acknowledge that surprise can happen.' If we face up to this fact and focus on maintaining (or in many cases improving) resilience, and considering a diverse range of management strategies, we are more likely to weather surprises. Moreover, increasingly we do understand the major characteristics of systems and activities that predispose the system to undergo a tipping point, allowing us to be more cautious in these systems/activity combinations and potentially focusing research on these areas. There is also evidence that monitoring in combination with adaptive management actions based on ecological knowledge can reduce the risk of surprise (Kelly et al., 2015; Hewitt & Thrush, 2019). In particular, we need to develop methods to assess risks associated with cumulative effects and understand the constraints to recovery.

All these factors, together with the progressive loss of large and long-lived organisms, decreases in habitat heterogeneity as well as in species diversity and function around the world have led to increasing interest in ecosystem-based management. Ecosystem-based management (Levin & Lubchenco, 2008; Marzloff et al., 2016; Thrush & Dayton, 2010) focusses on including the reality of connections within and between ecosystem components, managing cumulative stressors across a range of scales and trying to achieve sustainability for future generations, rather than optimising resource use. However, many of the tools needed to conduct ecosystem-based management effectively are still missing and will be a fruitful research topic for many years to come.

11.7 Close out

Increasingly in coastal soft-sediment systems, multiple stressors accumulating over time and, increasingly, large spatial scales are resulting in the loss of biodiversity and, in particular, of large and often long-lived taxa such as shellfish. Foundation species are often very sensitive to disturbance and their losses due to human impacts can profoundly change important functional relationships, including how biodiversity contributes to ecosystem functioning (see Chapter 10) and the resilience of the system to further change.

As ecologists we recognise the need to enhance biodiversity and redundancy in function and maintain spatial heterogeneity at multiple scales. This requires identifying key processes, and connectivity across scales of space, time and organisation. All this requires empirical data and a strong emphasis on studies which collect relevant information on the specific mechanisms by which specific stressors impact on species, while attempting to create

generalities that can be used to move across space, time and stressors.

References

Altieri A H and Diaz R J (2019). Dead Zones: Oxygen Depletion in Coastal Ecosystems. In: Sheppard C, ed. *World Seas: An Environmental Evaluation*, 2nd ed., Academic Press, San Diego, CA, pp. 453–73.

Ban S S, Graham N A and Connolly S R (2014). Evidence for Multiple Stressor Interactions and Effects on Coral Reefs. *Global Change Biology*, **20** (3), 681–97.

Beck M W, Brumbaugh R D, Airoldi L, Carranza A, Coen L D, Crawford C, Defeo O, Edgar G J, Hancock B, Kay M C, Lenihan H S, Luckenbach M W, Toropova C L, Zhang G and Guo X (2011). Oyster Reefs at Risk and Recommendations for Conservation, Restoration, and Management. *Bioscience*, **61** (2), 107–16.

Breitburg D, Levin L A, Oschlies A, Grégoire M, Chavez F P, Conley D J, Garçon V, Gilbert D, Gutiérrez D, Isensee K, Jacinto G S, Limburg K E, Montes I, Naqvi S W A, Pitcher G C, Rabalais N N, Roman M R, Rose K A, Seibel B A, Telszewski M, Yasuhara M and Zhang J (2018). Declining Oxygen in the Global Ocean and Coastal Waters. *Science*, **359**, eaam7240.

Buck K R, Rabalais N N, Bernhard J M and Barry J P (2012). Living Assemblages from the "Dead Zone" and Naturally Occurring Hypoxic Zones. In: Altenbach A V, Bernhard J M and Seckbach J, eds. *Anoxia: Evidence for Eukaryote Survival and Paleontological Strategies*, Springer, Dordrecht, pp. 343–52.

Carpenter S R, ed. (2003). *Regime Shifts in Lake Ecosystems: Pattern and Variation*, Ecology Institute, Oldendorf/Luhe.

Carson H S, Colbert S L, Kaylor M J and McDermid K J (2011). Small Plastic Debris Changes Water Movement and Heat Transfer through Beach Sediments. *Marine Pollution Bulletin*, **62** (8), 1708–13.

Carstensen J, Andersen J H, Gustafsson B G and Conley D J (2014). Deoxygenation of the Baltic Sea During the Last Century. *Proceedings of the National Academy of Sciences of the United States of America*, **111** (15), 5628–33.

Clark M R, Althaus F, Schlacher T A, Williams A, Bowden D A and Rowden A A (2016). The Impacts of Deep-Sea Fisheries on Benthic Communities: A Review. *ICES Journal of Marine Science: Journal du Conseil*, **73** (suppl 1), i51–i69.

Cluzard M, Kazmiruk T N, Kazmiruk V D and Bendell L (2015). Intertidal Concentrations of Microplastics and Their Influence on Ammonium Cycling as Related to the Shellfish Industry. *Archives of Environmental Contamination and Toxicology*, **69** (3), 310–19.

Cole M, Lindeque P, Halsband C and Galloway T S (2011). Microplastics as Contaminants in the Marine

Environment: A Review. *Marine Pollution Bulletin*, **62** (12), 2588–97.

Collie J S, Hall S J, Kaiser M J and Poiner I R (2000). A Quantitative Analysis of Fishing Impacts on Shelf-Sea Benthos. *Journal of Animal Ecology*, **69** (5), 785–98.

Conley D J, Carstensen J, Ærtebjerg G, Christensen P B, Dalsgaard T, Hansen J L and Josefson A B (2007). Long-Term Changes and Impacts of Hypoxia in Danish Coastal Waters. *Ecological Applications*, **17** (sp5), S165–84.

Conley D J, Carstensen J, Aigars J, Axe P, Bonsdorff E, Eremina T, Haahti B-M, Humborg C, Jonsson P, Kotta J, Lännegren C, Larsson U, Maximov A, Rodriguez Medina M, Lysiak-Pastuszak E, Remeikaitė-Nikienė N, Walve J, Wilhelms S and Zillén L (2011). Hypoxia Is Increasing in the Coastal Zone of the Baltic Sea. *Environmental Science & Technology*, **45** (16), 6777–83.

Cooper S R and Brush G S (1993). A 2,500-Year History of Anoxia and Eutrophication in Chesapeake Bay. *Estuaries*, **16** (3), 617–26.

Costanza R, d'Arge R, de Groot R, Farber S, Grasso M, Hannon B, Naeem S, Limburg K, Paruelo J, O'Neill R V, Raskin R, Sutton P and van den Belt M (1997). The Value of the World's Ecosystem Services and Natural Capital. *Nature*, **387**, 253–60.

Crain C M, Kroeker K and Halpern B S (2008). Interactive and Cumulative Effects of Multiple Human Stressors in Marine Systems. *Ecology Letters*, **11** (12), 1304–15.

Dayton P K, Tegner M J, Edwards P B and Riser K L (1998). Sliding Baselines, Ghosts, and Reduced Expectations in Kelp Forest Communities. *Ecological Applications*, **8** (2), 309–22.

Dayton P K, Thrush S F, Agardy T M and Hofman R J (1995). Environmental Effects of Fishing. *Aquatic Conservation: Marine and Freshwater Ecosystems*, **5**, 205–32.

de Juan S and Demestre M (2012). A Trawl Disturbance Indicator to Quantify Large Scale Fishing Impact on Benthic Ecosystems. *Ecological Indicators*, **18**, 183–90.

de Juan S and Hewitt J (2011). Relative Importance of Local Biotic and Environmental Factors versus Regional Factors in Driving Macrobenthic Species Richness in Intertidal Areas. *Marine Ecology Progress Series*, **423**, 117–29.

de Juan S, Hewitt J, Subidac M D and Thrush S (2018). Translating Ecological Integrity Terms into Operational Language to Inform Societies. *Journal of Environmental Management*, **228**, 319–27.

de Juan S, Lo Iacono C and Demestre M (2013). Benthic Habitat Characterisation of Soft-Bottom Continental Shelves: Integration of Acoustic Surveys, Benthic Samples and Trawling Disturbance Intensity. *Estuarine, Coastal and Shelf Science*, **117**, 199–209.

de Juan S, Thrush S F and Demestre M (2007). Functional Changes as Indicators of Trawling Disturbance on a Benthic Community Located in a Fishing Ground (NW Mediterranean Sea). *Marine Ecology Progress Series*, **334**, 117–29.

Diaz R J and Rosenberg R (2008). Spreading Dead Zones and Consequences for Marine Ecosystems. *Science*, **321** (5891), 926–9.

Eberhardt L L and Thomas J M (1991). Designing Environmental Field Studies. *Ecological Monographs*, **61**, 53–73.

GESAMP (Joint Group of Experts on the Scientific Aspects of Marine Environmental Protection) (1994). *Anthropogenic Influences on Sediment Discharge to the Coastal Zone and Environmental Consequences*, UNESCO-TOC, Paris.

Gladstone-Gallagher R V, Pilditch C A, Stephenson F and Thrush S F (2019). Linking Traits across Ecological Scales Determines Functional Resilience. *Trends in Ecology & Evolution*, **34**, 1080–91.

Green D S, Boots B, O'Connor N E and Thompson R (2017). Microplastics Affect the Ecological Functioning of an Important Biogenic Habitat. *Environmental Science & Technology*, **51** (1), 68–77.

Halpern B S, Selkoe K A, Micheli F and Kappel C V (2007). Evaluating and Ranking the Vulnerability of Global Marine Ecosystems to Anthropogenic Threats. *Conservation Biology*, **21** (5), 1301–15.

Hewawasam T, von Blanckenburg F, Schaller M and Kubik P (2003). Increase of Human over Natural Erosion Rates in Tropical Highlands Constrained by Cosmogenic Nuclides. *Geology*, **31** (7), 597–600.

Hewitt J, Thrush S, Lohrer A and Townsend M (2010). A Latent Threat to Biodiversity: Consequences of Small-Scale Heterogeneity Loss. *Biodiversity and Conservation*, **19**, 1315–23.

Hewitt J E, Lundquist C J and Ellis J (2018). Assessing Sensitivities of Marine Areas to Stressors Based on Biological Traits. *Conservation Biology*, **33**, 142–51.

Hewitt J E and Thrush S F (2019). Monitoring for Tipping Points in the Marine Environment. *Journal of Environmental Management*, **234**, 131–7.

Jackson J B, Kirby M X, Berger W H, Bjorndal K A, Botsford L W, Bourque B J, Bradbury R H, Cooke R, Erlandson J, Estes J A, Hughes T P, Kidwell S, Lange C B, Lenihan H S, Pandolfi J M, Peterson C H, Steneck R S, Tegner M J and Warner R R (2001). Historical Overfishing and the Recent Collapse of Coastal Ecosystems. *Science*, **293** (5530), 629–37.

Jackson J B C (2001). What Was Natural in the Coastal Oceans? *Proceedings of the National Academy of Sciences of the United States of America*, **98**, 5411–18.

Johnston E L and Roberts D A (2009). Contaminants Reduce the Richness and Evenness of Marine Communities: A Review and Meta-Analysis. *Environmental Pollution*, **157**, 1745–52.

Josefson A B, Norkko J and Norkko A (2012). Burial and Decomposition of Plant Pigments in Surface Sediments of the Baltic Sea: Role of Oxygen and Benthic Fauna. *Marine Ecology Progress Series*, **455**, 33–49.

Karlson K, Rosenberg R and Bonsdorff E (2002). Temporal and Spatial Large-Scale Effects of Eutrophication and Oxygen Deficiency on Benthic Fauna in Scandinavian and Baltic Waters—a Review. *Oceanography and Marine Biology: An Annual Review*, **40**, 427–89.

Kauppi L, Norkko A and Norkko J (2015). Large-Scale Species Invasion into a Low-Diversity System: Spatial and Temporal Distribution of the Invasive Polychaetes *Marenzelleria* Spp. in the Baltic Sea. *Biological Invasions*, **17**, 2055–74.

Kelly R P, Erickson A L, Mease L A, Battista W, Kittinger J N and Fujita R (2015). Embracing Thresholds for Better Environmental Management. *Philosophical Transactions of the Royal Society B: Biological Sciences*, **370** (1659), 2013.0276.

Law K L (2017). Plastics in the Marine Environment. *Annual Review of Marine Science*, **9**, 205–29.

Lefcheck J S, Marion S R and Orth R J (2017). Restored Eelgrass (*Zostera marina* L.) as a Refuge for Epifaunal Biodiversity in Mid-Western Atlantic Coastal Bays. *Estuaries and Coasts*, **40** (1), 200–12.

Lenihan H S, Micheli F, Shelton S W and Peterson C H (1999). The Influence of Multiple Environmental Stressors on Susceptibility to Parasites: An Experimental Determination with Oysters. *Limnology and Oceanography*, **44** (3), 910–24.

Levin S A and Lubchenco J (2008). Resilience, Robustness, and Marine Ecosystem-Based Management. *Bioscience*, **58** (1), 27–32.

Lipcius R N and Burke R P (2018). Successful Recruitment, Survival and Long-Term Persistence of Eastern Oyster and Hooked Mussel on a Subtidal, Artificial Restoration Reef System in Chesapeake Bay. *PloS One*, **13** (10), e0204329.

Lotze H K, Lenihan H S, Bourque B J, Bradbury R H, Cooke R G, Kay M C, Kidwell S M, Kirby M X, Peterson C H and Jackson J B (2006). Depletion, Degradation, and Recovery Potential of Estuaries and Coastal Seas. *Science*, **312** (5781), 1806–9.

McConnaughey R A, Mier K L and Dew C B (2000). An Examination of Chronic Trawling Effects on Soft-Bottom Benthos of the Eastern Bering Sea. *ICES Journal of Marine Science*, **57** (5), 1377–88.

McKnight D G (1969). A Recent, Possibly Catastrophic Burial in a Marine Molluscan Community. *New Zealand Journal of Marine and Freshwater Research*, **3**, 177–9.

Maltby L (1999). Studying Stress: The Importance of Organism-Level Responses. *Ecological Applications*, **9** (2), 431–40.

Mangano M C, Kaiser M J, Porporato E M, Lambert G I, Rinelli P and Spanò N (2014). Infaunal Community Responses to a Gradient of Trawling Disturbance and a Long-Term Fishery Exclusion Zone in the Southern Tyrrhenian Sea. *Continental Shelf Research*, **76**, 25–35.

Marzloff M P, Melbourne-Thomas J, Hamon K G, Hoshino E, Jennings S, Van Putten I E and Pecl G T (2016). Modelling Marine Community Responses to Climate-Driven Species Redistribution to Guide Monitoring and Adaptive Ecosystem-Based Management. *Global Change Biology*, **22** (7), 2462–74.

Maximov A (2011). Large-Scale Invasion of *Marenzelleria* Spp. (Polychaeta; Spionidae) in the Eastern Gulf of Finland, Baltic Sea. *Russian Journal of Biological Invasions*, **2** (1), 11–19.

Muntadas A, de Juan S and Demestre M (2015). Integrating the Provision of Ecosystem Services and Trawl Fisheries for the Management of the Marine Environment. *Science of the Total Environment*, **506**, 594–603.

Muntadas A, Demestre M, de Juan S and Frid C L (2014). Trawling Disturbance on Benthic Ecosystems and Consequences on Commercial Species: A Northwestern Mediterranean Case Study. *Scientia Marina*, **78** (S1), 53–65.

Norkko J, Reed D C, Timmermann K, Norkko A, Gustafsson B G, Bonsdorff E, Slomp C P, Carstensen J and Conley D J (2012). A Welcome Can of Worms? Hypoxia Mitigation by an Invasive Species. *Global Change Biology*, **18** (2), 422–34.

Oberle F K, Storlazzi C D and Hanebuth T J (2016). What a Drag: Quantifying the Global Impact of Chronic Bottom Trawling on Continental Shelf Sediment. *Journal of Marine Systems*, **159**, 109–19.

Officer C B, Biggs R B, Taft J L, Cronin L E, Tyler M A and Boynton W R (1984). Chesapeake Bay Anoxia: Origin, Development, and Significance. *Science*, **223** (4631), 22–7.

Oug E, Fleddum A, Rygg B and Olsgard F (2012). Biological Traits Analyses in the Study of Pollution Gradients and Ecological Functioning of Marine Soft Bottom Species Assemblages in a Fjord Ecosystem. *Journal of Experimental Marine Biology and Ecology*, **432**, 94–105.

Palanques A, Guillen J and Puig P (2001). Impact of Bottom Trawling on Water Turbidity and Muddy Sediment of an Unfished Continental Shelf. *Limnology and Oceanography*, **46** (5), 1100–10.

Pauly D (1995). Anecdotes and the Shifting Baseline Syndrome of Fisheries. *Trends in Ecology & Evolution*, **10**, 430.

Pearson T H and Rosenberg R (1978). Macrobenthic Succession in Relation to Organic Enrichment and Pollution of the Marine Environment. *Oceanography and Marine Biology: An Annual Review*, **16**, 229–311.

Peterson C H (1985). Patterns of Lagoonal Bivalve Mortality after Heavy Sedimentation and Their Paleoecological Significance. *Paleobiology*, **11**, 139–53.

Puig P, Canals M, Company J B, Martín J, Amblas D, Lastras G, Palanques A and Calafat A M (2012). Ploughing the Deep Sea Floor. *Nature*, **489** (7415), 286–9.

Rabalais N N, Turner R E and Wiseman W J J (2002). Gulf of Mexico Hypoxia, A.K.A. "The Dead Zone". *Annual Review of Ecology and Systematics*, **33**, 235–60.

Rick T C, Reeder-Myers L A, Hofman C A, Breitburg D, Lockwood R, Henkes G, Kellogg L, Lowery D, Luckenbach M W, Mann R, Ogburn M B, Southworth M, Wah J, Wesson J and Hines A H (2016). Millennial-Scale Sustainability of the Chesapeake Bay Native American Oyster Fishery. *Proceedings of the National Academy of Sciences of the United States of America*, **113** (23), 6568–73.

Robinson L A and Frid C L (2008). Historical Marine Ecology: Examining the Role of Fisheries in Changes in North Sea Benthos. *AMBIO: A Journal of the Human Environment*, **37** (5), 362–72.

Rothschild B J, Ault J S, Goulletquer P and Héral M (1994). Decline of the Chesapeake Bay Oyster Population: A Century of Habitat Destruction and Overfishing. *Marine Ecology Progress Series*, **111**, 29–39.

Schimel J, Balser T C and Wallenstein M (2007). Microbial Stress-Response Physiology and Its Implications for Ecosystem Function. *Ecology*, **88** (6), 1386–94.

Seabloom E W, Dobson A P and Stoms D M (2002). Extinction Rates under Nonrandom Patterns of Habitat Loss. *Proceedings of the National Academy of Sciences of the United States of America*, **99** (17), 11229–34.

Thrush S F and Dayton P (2010). What Can Ecology Contribute to Ecosystem-Based Management of Marine Fisheries? *Annual Reviews in Marine Science*, **2**, 419–41.

Thrush S F, Ellingsen K E and Davis K (2015). Implications of Fisheries Impacts to Seabed Biodiversity and Ecosystem-Based Management. *ICES Journal of Marine Science*, **73** (suppl 1), i44–i50.

Thrush S F, Gray J S, Hewitt J E and Ugland K I (2006). Predicting the Effects of Habitat Homogenization on Marine Biodiversity. *Ecological Applications*, **16**, 1636–42.

Thrush S F, Hewitt J E, Cummings V J, Ellis J I, Hatton C, Lohrer A and Norkko A (2004). Muddy Waters: Elevating Sediment Input to Coastal and Estuarine Habitats. *Frontiers in Ecology and Environment*, **2**, 299–306.

Thrush S F, Hewitt J E, Hickey C W and Kelly S (2008). Multiple Stressor Effects Identified from Species Abundance Distributions: Interactions between Urban Contaminants and Species Habitat Relationships *Journal of Experimental Marine Biology and Ecology*, **366**, 160–8.

Thrush S F, Lewis N, Le Heron R, Fisher K, Lundquist C J and Hewitt J E (2016). Addressing Surprise and Uncertain Futures in Marine Science, Marine Governance and Society. *Ecology and Society*, **21**, 44.

Tomimatsu H, Sasaki T, Kurokawa H, Bridle J R, Fontaine C, Kitano J, Stouffer D B, Vellend M, Bezemer T M and Fukami T (2013). Sustaining Ecosystem Functions in a Changing World: A Call for an Integrated Approach. *Journal of Applied Ecology*, **50** (5), 1124–30.

Tyler-Walters H, Rogers S I, Marshall C E and Hiscock K (2009). A Method to Assess the Sensitivity of Sedimentary Communities to Fishing Activities. *Aquatic Conservation: Marine and Freshwater Ecosystems*, **19** (3), 285–300.

Ugland K I and Gray J S (1982). Lognormal Distributions and the Concept of Community Equilibrium. *Oikos*, **39**, 171–8.

United Nations, Department of Economic and Social Affairs, Population Division (2017). World Population Prospects: The 2017 Revision, Key Findings and Advance Tables. *Working Paper No. ESA/P/WP/248*.

van Denderen P D, Hintzen N T, Rijnsdorp A D, Ruardij P and van Kooten T (2014). Habitat-Specific Effects of Fishing Disturbance on Benthic Species Richness in Marine Soft Sediments. *Ecosystems*, **17** (7), 1216–26.

Wang J, Tan Z, Peng J, Qiu Q and Li M (2016). The Behaviors of Microplastics in the Marine Environment. *Marine Environmental Research*, **113**, 7–17.

Whitlatch R B, Lohrer A M and Thrush S F (2001). Scale-Dependent Recovery of the Benthos: Effects of Larval and Post-Larval Stages. In: Aller J Y, Woodin S A and Aller R C, eds. *Organism-Sediment Interactions*, University of South Carolina, Columbia, SC, pp. 181–99.

Zacharias M A and Gregr E J (2005). Sensitivity and Vulnerability in Marine Environments: An Approach to Identifying Vulnerable Marine Areas. *Conservation Biology*, **19** (1), 86–97.

Zillén L and Conley D J (2010). Hypoxia and Cyanobacteria Blooms—Are They Really Natural Features of the Late Holocene History of the Baltic Sea? *Biogeosciences*, **7** (8), 2567–80.

CHAPTER 12

Climate change and seafloor ecology

12.1 Introduction

Climate change is one of the major environmental challenges of our time. Of all the chapters in this book, this one is likely to go out of date the fastest. In fact, in the time it has taken us to write the book, the predictions of climate change have shifted significantly and not in a good way. The accuracy and precision of physical predictions of climate change are important but the critical issue is how these changes will affect social-ecological systems. A large part of this very broad scale problem that is relevant to the ecology of marine sediments is: how does climate change affect the structure and function of seafloor ecosystems? We can consider climate change as an example of how soft-sediment benthic ecology can contribute to understanding and mitigating large-scale complex environmental problems. It's important to think of climate change as a cumulative and multiple stressor problem, not only because of the different aspects of climate change (e.g. changes in temperature, storm frequency, ocean acidification and sea level rise) but also because climate change will interact with many other stressors (e.g. warming of the water will reduce oxygen concentration exasperating the effects of eutrophication, see Figure 12.1).

There are numerous researchers working on climate change and looking for possible solutions and adaptations. Although a lot of research is needed, so too are syntheses and connections if we are to have effective actions to mitigate impacts and develop adaptation strategies. Thus, one common thread that currently runs through the literature is a call for more integrated approaches to connect disparate projects, although successful models of how we might do this have yet to emerge (Castree et al., 2014).

Researchers have studied climate change impacts from many different perspectives, currently with little integration across systems or scales of biological organisation. This is a classic example of the problem of studying the mechanistic detail of cause and effect relationships at one scale without considering the broader network of interactions in which the specifics of a particular study occur. This can limit our ability to understand the potential constraints, compensation and opportunity provided by change. As we think about climate change to coastal ecosystems, we need to consider both impacts, and the opportunities that maintaining or enhancing specific ecosystem functions may bring to, adaptation and mitigation strategies (see Chapter 13). Researching processes across different space and time scales will be important because we can expect climate effects to manifest differently in different places due to location, legacy effects, ecosystem type, biodiversity and the interactions with other stressors. Importantly, climate change involves rapid change for ecosystems. The frequency of marine heatwaves is increasing (Smale et al., 2019), but often the shallow areas that are vulnerable to rapid heating are poorly monitored.

Monitoring is an essential part of understanding *what* is happening to specific components of coastal ecosystems, but also in helping to understand *why* they are happening. Monitoring has been an essential tool in demonstrating the real-world changes that have been at the forefront of initiating the climate change research agenda. The ocean has absorbed about 30% of atmospheric CO_2 between

Ecology of Coastal Marine Sediments: Form, Function, and Change in the Anthropocene. Simon F. Thrush, Judi E. Hewitt, Conrad A. Pilditch and Alf Norkko,
Oxford University Press (2021). © Simon Thrush, Judi Hewitt, Conrad Pilditch, and Alf Norkko.
DOI: 10.1093/oso/9780198804765.003.0012

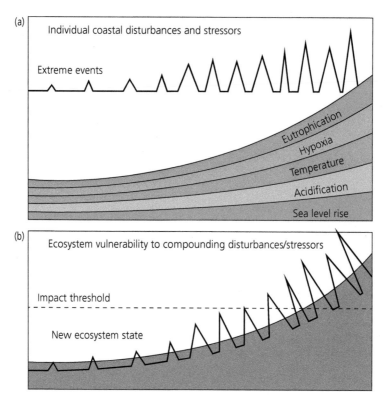

Figure 12.1 Coastal ecosystem disturbances, stressors and vulnerability. (A) Increasing air and water temperatures, water acidification, rates of sea-level rise, eutrophication, hypoxia, and frequency/magnitude of extreme storm surge events are among the primary threats to the ecology and hydro-biogeochemistry of coastal interfaces. (B) Although the resilience of coastal ecosystems is relatively unknown, it is likely that compounding disturbances and chronic stress will eventually exceed their impact threshold, resulting in widespread collapse of ecological function. Additional drivers of change not shown include land-use change, river impoundment, natural resource extraction, invasive species, droughts, floods and fires. From Ward N D, Megonigal J P, Bond-Lamberty B, Bailey V L, Butman D, Canuel E A, Diefenderfer H, Ganju N K, Goñi M A, Graham E B, Hopkinson C S, Khangaonkar T, Langley J A, McDowell N G, Myers-Pigg A N, Neumann R B, Osburn C L, Price R M, Rowland J, Sengupta A, Simard M, Thornton P E, Tzortziou M, Vargas R, Weisenhorn P B and Windham-Myers L 2020. Representing the Function and Sensitivity of Coastal Interfaces in Earth System Models. *Nature Communications*, **11**, 2458. Reprinted under Creative Commons Attribution 4.0 International (CC BY 4.0) licence.

the beginning of the industrial revolution and the mid-1990s, although this trend may not necessarily continue into the future (Gruber et al., 2019). Keeling's time series of atmospheric CO_2 concentrations from Mona Loa (Hawaii) were very powerful, even though the effects of burning fossil fuels had been predicted from first principles by the genius Arrhenius (Arrhenius, 1896). Monitoring will remain an important part of any scientific strategy to adapt to climate change if, as well as determining trends in environmental drivers, the monitoring is also designed to assess ecosystem responses. Both kinds of data are also vital in validating the accuracy of our predictions and success of management interventions. The value of integrating monitoring

data and model projections is illustrated by the work of Wethey et al. (2011, 2016) (see Figure 12.2), where shifts in the biogeography of polychaetes are driven by responses to temperature extremes (cold summers).

12.2 Ecosystem responses and drivers

Climate change, as with other forms of environmental disruption, results in soft-sediment species and habitats being either positively or negatively impacted. To avoid extinction, species will need to be resilient, adapt, move locally to exploit refugia or move over larger distances to colonise suitable regions. The opportunities for individual species

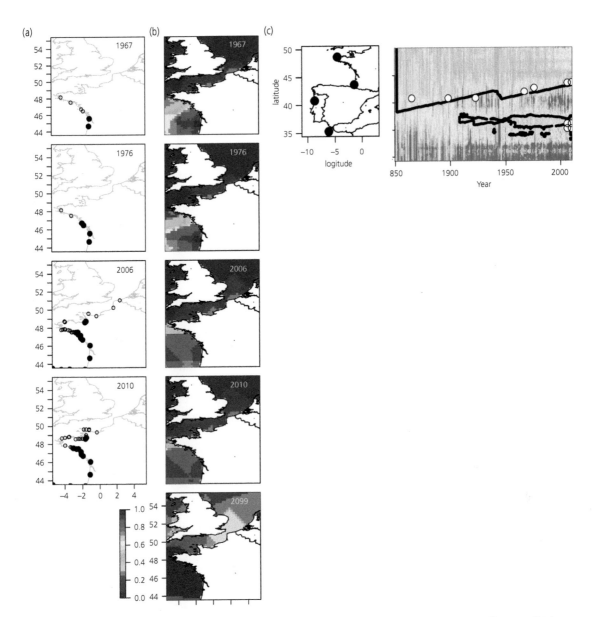

Figure 12.2 Changes in the distribution patterns of the onuphid polychaete *Diopatra biscayensis*, which is sensitive to cold summers (Wethey et al., 2016). Historical and recent data (a) are used to map *Diopatra*'s distribution in Europe. A combination of metapopulation models and coastal climate records and predictions are used to map *Diopatra*'s distribution. (b) The colour code represents the proportion of model runs that indicate *Diopatra* were present in a particular model cell. (c) shows changes in the northern geographic limit (thick line) driven by reproductive failure at low summer temperatures. Reprinted from *Journal of Experimental Marine Biology and Ecology*, **400**, Wethey D S, Woodin S A, Hiblish T J, Lima F P and Banrock P M, Response of Intertidal Populations to Climate: Effects of Extreme Events versus Long Term Change, 132–44., Copyright (2011), with permissions from Elsevier.

are likely to be different for species that are responding to average conditions and long-term trends associated with tracking a particular climate niche vs those that are responding to increased frequencies of episodic events. Nevertheless, individual species are recording distribution shifts (Crickenberger & Wethey, 2018; Hiddink et al., 2015), changes in behaviour (Monaco et al., 2016) and altered reproductive success (Herrera et al., 2019). But there remains much to learn about exactly how individual species will cope, let alone what the effects on higher levels of biological organisation will be.

Climate change has always occurred on our planet. These changes have been driven by episodic events such as meteorites, more gradual changes in earth systems, and the evolution of life on the planet, including the seafloor (Martin, 1996). But the anthropogenic effects of elevating atmospheric CO_2 and other greenhouse gas concentrations have to increased the rate and magnitude of change. On a global scale, one of the more recently obvious signs is an increase in marine heatwaves (Holbrook et al., 2019; Smale et al., 2019). These episodic phenomena are affecting the ecology of our planet, in ways and rates that cannot be informed by changes in the paleo record alone, due to the speed of change. There are still important uncertainties associated with changes in ocean currents, the degree of melting of ice caps, and the incorporation of feedback processes into predictive models (IPCC, 2019). In terms of the ecological consequences of these changes for seafloor ecology, it is important to also consider that immediate extinctions may be more likely driven by other human impacts and the ecological interactions that follow climate change (Paine, 1993; Clarke, 1993).

Coasts and estuaries are marine ecosystems at the forefront of climate change because of: complicated coastal topography and bathymetry affecting connectivity; land–coast interactions; strong coupling between seafloor and water column processes in shallow coastal waters; and the effects of multiple other stressors. Changes in storm frequency and intensity, and run-off from land will affect the intensity of upwelling events, tidal mixing and mixed layer depths—all factors likely to influence primary production (Doney et al., 2012; Hallegraeff, 2010). Increasing sea level rise may also result in the disappearance of many intertidal areas. The multiple ways that climate change effects will be manifest on different ecosystems mean we can expect strongly context-dependent changes around the world. Despite this, at present much published research continues to focus just on temperature and more recently ocean acidification. There is no clear hierarchy for prioritising how we might expect ecological systems to respond to change. Phenotypic plasticity, physiological tolerances, species sorting and environmental filtering, evolution, functional redundancy and habitat- and community-dependent effects on ecosystem function are all relevant lenses through which to view the ecology of change.

12.2.1 Temperature

Temperature is a primary driver of an organism's metabolic rate (Gillooly et al., 2001) and individual species have adapted to their environment to exhibit specific thermal tolerance ranges (Peck et al., 2014). Through many other chemical and physical changes, temperature has a wide range of effects on seafloor ecology and ecosystem function. For example, warming water increases the risk of hypoxia and changes in sediment biogeochemical processes, especially those coupled across aerobic–anaerobic interfaces. Temperature tolerances influence biogeographic distributions—this is termed a species' thermal niche. On a physiological level, the thermal tolerance of a species is linked to the oxygen requirements of its cells (Deutsch et al., 2015). Species that live in situations where temperature varies widely are likely to cope with temperature stress whereas species living under very stable temperature regimes, for example Antarctica, are much more susceptible to small changes. The thermal niche of a species may change associated with life stage or shifts in habitat utilisation but, locally, extreme events may result in rapid loss of a species if thermal tolerance levels are exceeded.

Broad-scale shifts in species distributions are often predicted from species distribution models (SDMs). These are built by relating the observed abundance, presence or biomass of species, or community attributes (e.g. diversity), to measured environmental factors or factors inferred from larger-scale physical models. There is an increasing depth of research into different ways of developing SDMs using various statistical models but typically SDMs focus on environmental factors alone. The challenge is to also incorporate ecological interactions (Dormann et al., 2012; Kraan et al., 2015). Linking these SDM approaches to physiological and ecological limits is an important task needed to increase our confidence in how species distributions may respond to climate change. Changes in mean temperature are not necessarily the most important effect of temperature rise; episodic high temperatures and desiccation stress are directly leading to mass-mortality events or impacts on organisms already stressed by disease or infection. This has been demonstrated by a shellfish die-off in Whangateau Harbour, New Zealand. Multiple late summer afternoons with extreme low tides and hot weather impacted the populations of *Austrovenus stutchburyi* that were already stressed by high infection with a coccidian parasite and a *Mycobacterium* infection. This combination triggered the mortality event that killed 60% of the intertidal population, with the large cockles (>30 mm shell length) more heavily impacted. Refugia for shellfish from the heat shock occurred in the shallow subtidal habitats so that recruitment back on the intertidal flats happened quickly. However, 10 years after the event, the density of shellfish has recovered but they have not reached the individual sizes that would be expected based on growth models (Tricklebank et al., 2020). These events can have long-term consequences when the organisms have important ecological roles in sediments and they are slow to recover.

Changes in the strength of seasonality associated with changes in temperature can also influence the timing of primary production and the nature of recruitment patterns: broadening the recruitment period of currently cold temperate species may offer increased opportunity for dispersal to new locations with more amenable conditions. However, coastal habitat fragmentation and isolation associated with changes especially in estuarine and intertidal habitats or changes in the oceanographic condition that support dispersal can generate barriers to species movement (Lipcius et al., 2008). The rapid reclamation and development of seawalls in China has resulted in loss of coastal habitat at a rate

of 40,000 ha year^{-1} from 2006 to 2010, isolating wetland and adjacent soft-sediment habitats with a loss of ecosystem services (Ma et al., 2014, 2019; Ouyang et al., 2018).

12.2.2 Ocean acidification

Another primary physico-chemical driver of climate change is ocean acidification (OA; Doney et al., 2009). Taking up about 1 M tonnes of CO_2 per hour (Sabine et al., 2004), the ocean absorbs about 25% of the CO_2 emitted into the atmosphere. This has increased the concentration of CO_2, H^+ and HCO_{3-}, shifted the acid–base balance and decreased the concentration of CO_3^{2-} and $CaCO_3$ saturation state (Gattuso et al., 1998). This acidification of seawater has direct effects on some coastal species (Portner, 2008; Widdicombe & Spicer, 2008). The effects are often negative on animals such as molluscs that rely on carbonate for shell formation, whereas for other species the effects are less clear cut. For some, such as seagrass, the elevated CO_2 and the removal of grazers can be beneficial (Kroeker et al., 2013; Waldbusser & Salisbury, 2014).

Most of the large-scale predictions of OA effects do not involve the interaction between the sediments and the water column, yet this is clearly critical for coastal and estuarine ecosystems. pH naturally changes rapidly from the overlying water and through the sediment profile, with often a change of about 1 pH unit occurring within the top 1–2 cm of sediments. This level of change is well in excess of the current climate change scenarios. This does not mean that OA effects are irrelevant to soft-sediment habitats but it underscores the importance of system understanding in order to ask appropriate questions and target solutions to management.

12.3 Improving our understanding of responses

The sequence and magnitude of specific changes, e.g. warming vs sea level rise, will influence biodiversity, ecosystem function and ecosystem services. With these changes occurring over different space and time scales, how can we generate empirical information to learn about the responses to climate change? Given the complex interplay of factors associated with climate change, good empirical data are essential to inform (parametrise and appropriately wire up) and test models and to assess if the management actions that we take are effective at achieving the outcomes we desire.

Time series data are very useful in helping us to understand how climate change effects could manifest in the future, especially where the data include variability associated with multi-year variability, such as that generated by the ENSO or NAO (see Chapter 8). Teasing apart variability in the abundance of sandflat macrofauna associated with large-scale climate forcing (ENSO) demonstrated how different scales of physical and biological processes influenced the temporal variability of monitored species (Hewitt & Thrush, 2009). Understanding the combination of factors that have affected ecosystem dynamics in the past can also help us better understand what changes may happen in the future.

Latitudinal gradient studies allow us to assess changes in species, communities or ecosystems along gradients that relate to environmental variables that are predicted to change with climate change. This approach was used to look at changes along the Victorialand and McMurdo Sound coast of the Ross Sea, Antarctica to assess community composition and foodweb structure in coastal benthic communities (Cummings et al., 2006, 2018; Norkko et al., 2007; Thrush et al., 2006). Coastal fast sea ice has a profound influence on seafloor ecology. Ice of the order of 1–4 m thick severely restricts light that can be further attenuated by drifting snow and dust. This limits local primary production by macroalgae, phytoplankton, under-ice algae and microphytobenthos, although effects on other trophic levels may be limited if currents allow primary food resources to be advected under the ice from open water. Sampling in locations that differ in the potential for advection of open water production, sea ice thickness and persistence allows us a glimpse of ecological relationships that are likely part of the climate change scenario for this region. These studies revealed changes in diversity, the depth distribution of primary producers, the density and size of large benthic species such as scallops and urchins, and changes in the utilisation of different food resources with increasing access to high-quality plant material.

What is truly amazing in this system is the role of episodic events that link to changes in sea ice—these can create ecological legacies for decades. Early diving studies by Paul Dayton investigated the growth of large glass sponges on the seafloor in McMurdo Sound. Growth rates were low for two to three decades but recently ice break-out occurred with sponges recruiting and growing rapidly (Dayton et al., 2016). These kinds of studies that extend across time and space and involve important natural history insights can help us understand the interactions between different physical and biological processes that manifest as climate change (Dayton et al., 2019).

Given the geographic biases that often exist in soft-sediment research we need to be careful how we generalise information about soft-sediment structure and function. Climate change brings this problem into sharp focus but it also offers a tremendous opportunity for large-scale experiments that test theory and functionality across conditions and therefore provide insight into likely climate change effects.

Studies of extreme events relate closely to disturbance-recovery phenomena (see Chapter 3). Whether it is heatwaves or storms, for specific coastal regions the frequency and intensity are expected to increase, changing the magnitude and extent of disturbance to the seafloor (Burrows et al., 2011; Holbrook et al., 2019; Paerl et al., 2001; Smale et al., 2019; Sukumaran et al., 2016). But impacts on the benthos will not only be generated by these direct stressors. On land, intense rainfall can increase the potential for sediment run-off leading to major deposits of terrestrial sediment in the coastal zone. This can have immediate effects—smothering benthic communities, making sedimentary habitats unfavourable for colonising species—as well as longer-term effects on sediment grain size composition and suspended sediment concentrations. Understanding these types of effects that are already impacting on soft-sediment biodiversity and ecosystem function can help us assess the risk of changes to coastal regions as climate and land–coast interaction models improve. This information is of extreme value in terms of risk assessment and hazard mitigation associated with climate change.

The many different environmental and biological changes associated with climate change really highlight the importance of implementing a cumulative effects or multiple stressor approach to research (Figure 12.3). There is the potential for multiple stressor effects to occur as a result of the interaction of different elements of climate change—but it is likely that climate change will make studying or managing any other environmental stressor a multiple stressor problem. To do this effectively, rather than simply becoming bewildered by the number of possible connections, it is critical we understand how sedimentary ecosystems function and the different components are connected. We also need to think seriously about space and time scales of change. These insights help us to define questions and focus our research to make it tractable (Box 12.1).

Justifiably, research on the effects of climate change on benthic communities has expanded tremendously over the last decade. What can we learn from an area of such rapid expansion? Importantly, that there is tremendous value in linking across different aspects of response, e.g. connecting physiological limits to ecological changes and the role of species in community and ecosystem function. Many of the ecological effects highlight how the combinations of environmental factors and current ecological status drive differences in responses from place to place and region to region (Woodin et al., 2013). This means that the effects of climate change may also vary over time as not all the factors involved will affect the benthic community at the same rate. For instance, temperature effects might be initially positive but then, when combined with subsequent OA effects, become negative. Climate change is often driven by what we consider today to be extreme events and the increasing frequency of these episodic events can change the ecosystem in very different ways from a more progressive and linear change. Non-linear change in ecosystems is

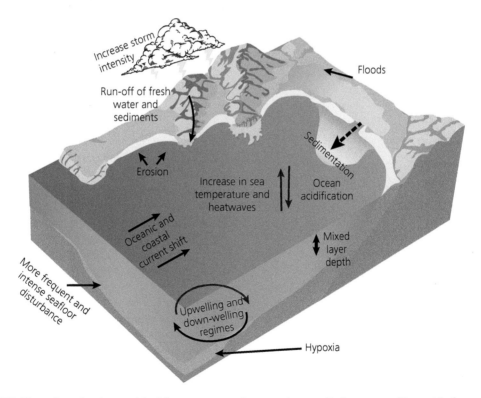

Figure 12.3 Climate change has the potential to influence many coastal processes, in terms of both average conditions and the frequency and intensity of extreme events. The relative importance of these factors on soft-sediment ecology will vary geographically with benthic populations, communities and ecosystem functions.

Box 12.1 Considering primary production in the context of climate change reveals different and interacting processes

Microphytes in sediments are expected to benefit from elevated CO_2 through enhanced photosynthesis, but changes in photosynthetically active radiation (PAR) and UVB are potentially limiting (Gao et al., 2019). Meanwhile, elevated temperature could increase grazing pressure and rates of microbial breakdown (Hicks et al., 2018). These counteracting drivers will play out differently from place to place. If coastal water becomes more turbid, associated with increased sediment run-off or resuspension with increasing storminess, the potential for production on the seafloor will decrease. Sea-level rise further complicates the situation. In an experimental study O'Meara et al. (2017) illustrated how the combination of adding nutrients, fine sediments and elevated water levels interact to drive increased variability and decoupling of ecosystem processes associated with nutrient regeneration in the sediments, shifting productivity from microphytobenthos to the water column. This study used experimental treatments that changed environmental conditions in a consistent fashion. However, extreme events might have quite different effects on primary producers, for example by generating the right conditions to stimulate phytoplankton blooms (Cloern et al., 2016). The quality of phytoplankton blooms also matters and the responses to a future climate are also influenced by the legacy of the past. Using models of the Baltic Sea, the ecosystem's slow recovery from eutrophication along with climate projections has indicated intensified recycling of organic matter in the water column, due to changes in species composition in the plankton, a weakened benthic–pelagic coupling and less benthic macrofaunal production (Ehrnsten et al., 2020).

likely as interactions and feedbacks change. Climate and oceanographic changes are implicated in contributing to the crossing of many ecological tipping points (Selkoe et al., 2015). This provides a vast opportunity for research that addresses the nature of change and chance in seafloor ecology.

12.4 Mitigation

Apart from understanding the changes in structure and function brought about by climate change, it is increasingly important we focus on how our understanding can be used to conserve and manage seafloor habitats in ways that mitigate effects or allow adaptation to climate change. It could be argued that hidden seafloor ecosystem services are already contributing to ameliorating change and therefore we do not need to focus attention here. However, we may be able to stop degradation and enhance service delivery if we make the right interventions. This raises two big problems.

Firstly, we need multiple strategies and actions to address complex and multifaceted problems like climate change, and we need to be very cautious about simple silver-bullet fixes. For example, the long-term storage of carbon on the seafloor is simplistically about the balance between primary production, consumption and respiration. This means that anoxic sediments, as found in the world's dead zones, are good places to store carbon. Although this is correct, the biogeochemistry of these anaerobic sediments means that they release methane and other climate warming gases. The excess nutrients that are generated cause other problems such as harmful algal blooms. Also these anaerobic sediments are devoid of organisms that require oxygen. This raises deep questions about human–nature relationships because in this case multiple functions are not working that support other ecosystem services.

The second problem follows: the disparity of world views between different sciences. There is a massive opportunity to develop the science and understanding necessary to bridge these gaps and generate a more holistic approach (Middelburg, 2018; Snelgrove et al., 2017). We can caricature this situation as encompassing a geochemical world view which focusses on chemical processes in a homogeneous environment and an ecological world view that focusses on heterogeneity and complexity. The big challenge is to scale up our understanding of functions and processes without losing the ecological roles of diversity, heterogeneity and connectivity or missing the importance of extreme events and the co-occurrence of events.

Many of the ecological effects that could contribute to mitigating the effects of climate change involve species or habitats that are already substantively impacted by human activity. For example, oyster reefs on the east coast of the USA can act as sources or sinks of carbon depending on reef age and environmental setting (Fodrie et al., 2017). There is evidence that healthy coastal plant communities such as seagrass and seaweeds can locally ameliorate lowering ocean pH and carbonate chemistry (Duarte et al., 2013; Raven, 2018). Seaweeds are often highly productive plants that can protect their tissues from herbivory. These plants can be torn from the reefs by waves and, on narrow continental shelfs and adjacent to canyons, get subducted into the deep sea (Vetter & Dayton, 1999). Plant material that is not consumed on the way to the deep sea can contribute to carbon storage for a long time.

In sediments, rates of carbon consumption decrease rapidly with depth into the sediment (O'Meara et al., 2018) and this can broaden the scope of carbon storage (Queirós et al., 2019). Microphytobenthos are also very productive plants, less refractile than kelps; they can be found in clear waters from the intertidal flats to the shelf break and are estimated to produce about 500 million tonnes of organic carbon annually (Cahoon, 1999; Hope et al., 2020). This is a preferred food resource for many benthic species where feeding and bioturbation subduct material deeper into the sediments, but we have yet to work out what contribution this makes to carbon storage.

Managing human activities such as trawling and dredging that disturb and resuspend seafloor sediments could perhaps enhance carbon storage on our highly productive coastal and shelf ecosystems and limit nutrient release from sediments. Conversely, they could resuspend fine organic-rich sediments facilitating transport to the deep sea (Churchill, 1989; Martín et al., 2014; Oberle et al., 2016a, 2016b; Pilskaln et al., 1998). These options have yet to be considered in the suite of solutions to address climate change. More emphasis should perhaps be directed towards understanding carbon stocks and their turnover rates and the relationship to biodiversity.

As discussed in Chapter 1, pH decreases with organic load to the sediment, and this means that in coastal systems OA effects are currently often more tightly coupled to eutrophication than to global climate change effects (Cai et al., 2011). Controlling nutrient inputs and eutrophication in the coastal zones will therefore improve the capacity of these ecosystems to respond to climate change-driven OA effects. These local changes in the OA status of coastal waters highlight the importance of considering climate change impacts in the context of multiple stressor effects where proactive action to restore diversity, ecosystem function and heterogeneity may ameliorate local effects and provide refugia from climate-related stresses. We must face facts—there is not going to be a one-size-fits-all strategy to deal with climate change (Strong et al., 2014).

The potential to work with nature in soft-engineering solutions to limit the effects of storm surge and sea level rise will also be very location specific (Temmerman et al., 2013). These approaches use natural habitats such as salt marshes, mangroves or shellfish beds to extract energy from waves and storm surges, reducing flooding risk, while simultaneously providing ecological benefits such as habitat creation or water filtering (Morris et al., 2018). Such approaches may not work in situations where the physical conditions are too extreme or where there is no potential for retreat of these fringing habitats because of coastal squeeze. However, nature-based flood defences can be substantially cheaper than conventional methods (Jones et al., 2012).

12.5 Close out

Climate change highlights the importance of taking a different view of how we try to manage coastal ecosystems to enhance their resilience, mitigate impacts where possible and fully recognise the benefits we derive from marine ecosystems. It's not all bad news; although locally we may not be able to strongly affect climate change impacts, we could limit the impacts of eutrophication and input of contaminants into the coastal zone that worsen the ecological consequences.

Future change is inevitable, with the magnitude related to how quickly we can transition our societies and our economies to acknowledging interactions with nature. Given the potential for major environmental change and shifts in the magnitude and nature of ecosystem service provision associated with climate change, one of the biggest changes to management will be to focus on building resilience rather than maximising exploitation. Where we can, we will need to minimise other impacts on coastal ecosystems and combine this with adaptation strategies for coping with climate change.

These challenges open up new, future-focussed and more inclusive ways of managing coastal ecosystems that address multiple uses, multiple stressors and link science to society to environment in ways that recognise complexity and uncertainty (Castree et al., 2014; Craig, 2012; Stirling, 2010; Thrush et al., 2016). Given that climate change is unprecedented, the required management actions will be as well and so adaptive management, learning-by-doing, will require greater integration of scientific methods, technologies and knowledge. We will need to be part of a greater engagement with society that facilitates collaborative decision making.

References

Arrhenius S (1896). On the Influence of Carbonic Acid in the Air upon the Temperature of the Ground. *Philosophical Magazine and Journal of Science Series*, **5**, 237–76.

Burrows M T, Schoeman D S, Buckley L B, Moore P, Poloczanska E S, Brander K M, Brown C, Bruno J F, Duarte C M, Halpern B S, Holding J, Kappel C V, Kiessling W, O'Connor M I, Pandolfi J M, Parmesan C, Schwing F B, Sydeman W J and Richardson A J (2011). The Pace of Shifting Climate in Marine and Terrestrial Ecosystems. *Science*, **334** (6056), 652–5.

Cahoon L B (1999). The Role of Benthic Microalgae in Neritic Ecosystems. *Oceanography and Marine Biology*, **37**, 47–86.

Cai W J, Hu X, Huang W J, Murrell M C, Lehrter J C, Lohrenz S E, Chou W C, Zhai W, Hollibaugh J T, Wang Y, Zhao P, Guo X, Gundersen K, Dai M and Gong G C (2011). Acidification of Subsurface Coastal Waters Enhanced by Eutrophication. *Nature Geoscience*, **4** (11), 766–70.

Castree N, Adams W M, Barry J, Brockington D, Büscher B, Corbera E, Demeritt D, Duffy R, Felt U, Neves K, Newell P, Pellizzoni L, Rigby K, Robbins P, Robin L, Rose D B, Ross A, Schlosberg D, Sörlin S, West P, Whitehead M and Wynne B (2014). Changing the Intellectual Climate. *Nature Climate Change*, **4**, 763–8.

Churchill J H (1989). The Effect of Commercial Trawling on Sediment Resuspension and Transport over the Middle Atlantic Bight Continental Shelf. *Continental Shelf Research*, **9**, 841–64.

Clarke A (1993). Temperature and Extinction in the Sea: A Physiologist's View. *Paleobiology*, **19**, 499–518.

Cloern J E, Abreu P C, Carstensen J, Chauvaud L, Elmgren R, Grall J, Greening H, Johansson J O R, Kahru M, Sherwood E T, Xu J and Yin K (2016). Human Activities and Climate Variability Drive Fast-Paced Change across the World's Estuarine–Coastal Ecosystems. *Global Change Biology*, **22** (2), 513–29.

Craig R K (2012). Ocean Governance for the 21st Century: Making Marine Zoning Climate Change Adaptable. *Harvard Environmental Law Review*, **36** (2), 305–50.

Crickenberger S and Wethey D S (2018). Annual Temperature Variation as a Time Machine to Understand the Effects of Long-Term Climate Change on a Poleward Range Shift. *Global Change Biology*, **24** (8), 3804–19.

Cummings V J, Hewitt J E, Thrush S F, Marriott P M, Halliday N J and Norkko A (2018). Linking Ross Sea Coastal Benthic Communities to Environmental Conditions: Documenting Baselines in a Spatially Variable and Changing World. *Frontiers in Marine Science*, **5**, 232.

Cummings V J, Thrush S F, Norkko A, Andrew N L, Hewitt J E, Funnell G A and Schwarz A-M (2006). Accounting for Local Scale Variability in Macrobenthos in Developing Assessments of Latitudinal Trends in the South Western Ross Sea. *Antarctic Science*, **18**, 633–44.

Dayton P, Jarrell S, Kim S, Thrush S, Hammerstrom K, Slattery M and Parnell E (2016). Surprising Episodic Recruitment and Growth of Antarctic Sponges: Implications for Ecological Resilience. *Journal of Experimental Marine Biology and Ecology*, **482**, 38–55.

Dayton P K, Jarrell S C, Kim S, Ed Parnell P, Thrush S F, Hammerstrom K and Leichter J J (2019). Benthic Responses to an Antarctic Regime Shift: Food Particle Size and Recruitment Biology. *Ecological Applications*, **29** (1), e01823.

Deutsch C, Ferrel A, Seibel B, Pörtner H O and Huey R B (2015). Climate Change Tightens a Metabolic Constraint on Marine Habitats. *Science*, **348** (6239), 1132–5.

Doney S C, Fabry V J, Feely R A and Kleypas J A (2009). Ocean Acidification: The Other CO_2 Problem. *Annual Review of Marine Science*, **1**, 169–92.

Doney S C, Ruckelshaus M, Emmett Duffy J, Barry J P, Chan F, English C A, Galindo H M, Grebmeier J M, Hollowed A B, Knowlton N, Polovina J, Rabalais N N, Sydeman W J and Talley L D (2012). Climate Change Impacts on Marine Ecosystems. *Annual Review of Marine Science*, **4**, 11–37.

Dormann C F, Schymanski S J, Cabral J, Chuine I, Graham C, Hartig F, Kearney M, Morin X, Römermann C, Schröder B and Singer A (2012). Correlation and Process in Species Distribution Models: Bridging a Dichotomy. *Journal of Biogeography*, **39** (12), 2119–31.

Duarte C M, Hendriks I E, Moore T S, Olsen Y S, Steckbauer A, Ramajo L, Carstensen J, Trotter J A and McCulloch M J E (2013). Is Ocean Acidification an Open-Ocean Syndrome? Understanding Anthropogenic Impacts on Seawater pH. *Estuaries and Coasts*, **36** (2), 221–36.

Ehrnsten E, Norkko A, Müller-Karulis B, Gustafsson E and Gustafsson B G (2020). The Meagre Future of Benthic Fauna in a Coastal Sea—Benthic Responses to Recovery from Eutrophication in a Changing Climate. *Global Change Biology*, **26** (4), 2235–50.

Fodrie F J, Rodriguez A B, Gittman R K, Grabowski J H, Lindquist N L, Peterson C H, Piehler M F and Ridge J T (2017). Oyster Reefs as Carbon Sources and Sinks. *Proceedings of the Royal Society B: Biological Sciences*, **284** (1859), 2017.0891.

Gao K, Beardall J, Häder D-P, Hall-Spencer J M, Gao G and Hutchins D A (2019). Effects of Ocean Acidification on Marine Photosynthetic Organisms under the Concurrent Influences of Warming, UV Radiation, and Deoxygenation. *Frontiers in Marine Science*, **6**, 322.

Gattuso J P, Frankignoulle M and Wollast R (1998). Carbon and Carbonate Metabolism in Coastal Aquatic Ecosystems. *Annual Review of Ecology and Systematics*, **29**, 405–34.

Gillooly J F, Brown J H, West G B, Savage V M and Charnov E L (2001). Effects of Size and Temperature on Metabolic Rate. *Science*, **293** (5538), 2248–51.

Gruber N, Clement D, Carter B R, Feely R A, van Heuven S, Hoppema M, Ishii M, Key R M, Kozyr A, Lauvset S K, Monaco C L, Mathis J T, Murata A, Olsen A, Perez F F, Sabine C L, Tanhua T and Wanninkhof R (2019). The Oceanic Sink for Anthropogenic CO_2 from 1994 to 2007. *Science*, **363** (6432), 1193–9.

Hallegraeff G M (2010). Ocean Climate Change, Phytoplankton Community Responses, and Harmful Algal Blooms: A Formidable Predictive Challenge. *Journal of Phycology*, **46** (2), 220–35.

Herrera M, Wethey D S, Vázquez E and Macho G (2019). Climate Change Implications for Reproductive Success: Temperature Effect on Penis Development in the Barnacle *Semibalanus balanoides*. *Marine Ecology Progress Series*, **610**, 109–23.

Hewitt J E and Thrush S F (2009). Reconciling the Influence of Global Climate Phenomena on Macrofaunal Temporal Dynamics at a Variety of Spatial Scales. *Global Change Biology*, **15**, 1911–29.

Hicks N, Liu X, Gregory R, Kenny J, Lucaci A, Lenzi L, Paterson D M and Duncan K R (2018). Temperature Driven Changes in Benthic Bacterial Diversity Influences Biogeochemical Cycling in Coastal Sediments. *Frontiers in Microbiology*, **9**, 1730.

Hiddink J G, Burrows M T and García Molinos J (2015). Temperature Tracking by North Sea Benthic Invertebrates in Response to Climate Change. *Global Change Biology*, **21** (1), 117–29.

Holbrook N J, Scannell H A, Sen Gupta A, Benthuysen J A, Feng M, Oliver E C J, Alexander L V, Burrows M T, Donat M G, Hobday A J, Moore P J, Perkins-Kirkpatrick S E, Smale D A, Straub S C and Wernberg T (2019). A Global Assessment of Marine Heatwaves and Their Drivers. *Nature Communications*, **10** (1), 2624.

Hope J A, Paterson D M and Thrush S (2020). The Role of Microphytobenthos in Soft-Sediment Ecological Networks and Their Contribution to the Delivery of Multiple Ecosystem Services. *Journal of Ecology*, **108** (3), 815–30.

IPCC (2019). *The Ocean and Cryosphere in a Changing Climate: Summary for Policymakers*, Intergovernmental Panel on Climate Change, Geneva.

Jones H P, Hole D G and Zavaleta E S (2012). Harnessing Nature to Help People Adapt to Climate Change. *Nature Climate Change*, **2** (7), 504–9.

Kraan C, Dormann C F, Greenfield B L and Thrush S F (2015). Cross-Scale Variation in Biodiversity-Environment Links Illustrated by Coastal Sandflat Communities. *PLoS One*, **10** (11), e0142411.

Kroeker K J, Kordas R L, Crim R, Hendriks I E, Ramajo L, Singh G S, Duarte C M and Gattuso J P (2013). Impacts of Ocean Acidification on Marine Organisms: Quantifying Sensitivities and Interaction with Warming. *Global Change Biology*, **19** (6), 1884–96.

Lipcius R N, Eggleston D B, Schreiber S J, Seitz R D, Shen J, Sisson M, Stockhausen W T and Wang H V (2008). Importance of Metapopulation Connectivity to Restocking and Restoration of Marine Species. *Reviews in Fisheries Science*, **16** (1–3), 101–10.

Ma T, Li X, Bai J and Cui B (2019). Habitat Modification in Relation to Coastal Reclamation and Its Impacts on Waterbirds along China's Coast. *Global Ecology and Conservation*, **17**, e00585.

Ma Z, Melville D S, Liu J, Chen Y, Yang H, Ren W, Zhang Z, Piersma T and Li B (2014). Rethinking China's New Great Wall. *Science*, **346** (6212), 912–14.

Martín J, Puig P, Palanques A and Giamportone A (2014). Commercial Bottom Trawling as a Driver of Sediment Dynamics and Deep Seascape Evolution in the Anthropocene. *Anthropocene*, **7**, 1–15.

Martin R E (1996). Secular Increase in Nutrient Levels through the Phanerozoic: Implications for Productivity, Biomass and Diversity of the Marine Biosphere. *Palaios*, **11**, 209–19.

Middelburg J J (2018). Reviews and Syntheses: To the Bottom of Carbon Processing at the Seafloor. *Biogeosciences*, **15** (2), 413–27.

Monaco C J, Wethey D S and Helmuth B (2016). Thermal Sensitivity and the Role of Behavior in Driving an

Intertidal Predator–Prey Interaction. *Ecological Monographs*, **86** (4), 429–47.

Morris R L, Konlechner T M, Ghisalberti M and Swearer S (2018). From Grey to Green: Efficacy of Eco-Engineering Solutions for Nature-Based Coastal Defence. *Global Change Biology*, **24** (5), 1827–42.

Norkko A, Thrush S F, Cummings V J, Gibbs M M, Andrew N L, Norkko J and Schwarz A-M (2007). Trophic Structure of Coastal Antarctic Food Webs Associated with Changes in Sea Ice and Food Supply. *Ecology*, **88**, 2810–20.

O'Meara T, Gibbs E and Thrush S F (2018). Rapid Organic Matter Assay of Organic Matter Degradation across Depth Gradients within Marine Sediments. *Methods in Ecology and Evolution*, **9** (2), 245–53.

O'Meara T A, Hillman J R and Thrush S F (2017). Rising Tides, Cumulative Impacts and Cascading Changes to Estuarine Ecosystem Functions. *Science Reports*, **7**, 10218.

Oberle F K J, Storlazzi C D and Hanebuth T J J (2016a). What a Drag: Quantifying the Global Impact of Chronic Bottom Trawling on Continental Shelf Sediment. *Journal of Marine Systems*, **159**, 109–19.

Oberle F K J, Swarzenski P W, Reddy C M, Nelson R K, Baasch B and Hanebuth T J J (2016b). Deciphering the Lithological Consequences of Bottom Trawling to Sedimentary Habitats on the Shelf. *Journal of Marine Systems*, **159**, 120–31.

Ouyang X, Lee S Y, Connolly R M and Kainz M J (2018). Spatially-Explicit Valuation of Coastal Wetlands for Cyclone Mitigation in Australia and China. *Scientific Reports*, **8** (1), 3035.

Paerl H W, Bales J D, Ausley L W, Buzzelli C P, Crowder L B, Eby L A, Fear J M, Go M, Peierls B L, Richardson T L and Ramus J S (2001). Ecosystem Impacts of Three Sequential Hurricanes (Dennis, Floyd, and Irene) on the United States' Largest Lagoonal Estuary, Pamlico Sound, NC. *Proceedings of the National Academy of Sciences of the United States of America*, **98** (10), 5655–60.

Paine R T (1993). A Salty and Salutary Perspective on Global Change. In: Kareiva P M, Kingsolver J G and Huey R B, eds. *Biotic Interactions and Global Change*, Sinauer, Sunderland, MA, pp. 347–55.

Peck L S, Morley S A, Richard J and Clark M S (2014). Acclimation and Thermal Tolerance in Antarctic Marine Ectotherms. *The Journal of Experimental Biology*, **217** (1), 16–22.

Pilskaln C H, Churchill J H and Mayer L M (1998). Resuspension of Sediments by Bottom Trawling in the Gulf of Maine and Potential Geochemical Consequences. *Conservation Biology*, **12**, 1223–4.

Portner H O (2008). Ecosystem Effects of Ocean Acidification in Times of Ocean Warming: A Physiologist's View. *Marine Ecology Progress Series*, **373**, 203–17.

Queirós A M, Stephens N, Widdicombe S, Tait K, McCoy S J, Ingels J, Rühl S, Airs R, Beesley A, Carnovale G, Cazenave P, Dashfield S, Hua E, Jones M, Lindeque P, McNeill C L, Nunes J, Parry H, Pascoe C, Widdicombe C, Smyth T, Atkinson A, Krause-Jensen D and Somerfield P J (2019). Connected Macroalgal-Sediment Systems: Blue Carbon and Food Webs in the Deep Coastal Ocean. *Ecological Monographs*, **89** (3), e01366.

Raven J (2018). Blue Carbon: Past, Present and Future, with Emphasis on Macroalgae. *Biology Letters*, **14** (10), 2018.0336.

Sabine C L, Feely R A, Gruber N, Key R M, Lee K, Bullister J L, Wanninkhof R, Wong C S, Wallace D W, Tilbrook B, Millero F J, Peng T H, Kozyr A, Ono T and Rios A F (2004). The Oceanic Sink for Anthropogenic CO_2. *Science*, **305** (5682), 367–71.

Selkoe K A, Blenckner T, Caldwell M R, Crowder L B, Erickson A L, Essington T E, Estes J A, Fujita R M, Halpern B S, Hunsicker M E, Kappel C V, Kelly R P, Kittinger J N, Levin P S, Lynham J M, Mach M E, Martone R G, Mease L A, Salomon A K, Samhouri J F, Scarborough C, Stier A C, White C and Zedler J (2015). Principles for Managing Marine Ecosystems Prone to Tipping Points. *Ecosystem Health and Sustainability*, **1** (5), 1–18.

Smale D A, Wernberg T, Oliver E C J, Thomsen M, Harvey B P, Straub S C, Burrows M T, Alexander L V, Benthuysen J A, Donat M G, Feng M, Hobday A J, Holbrook N J, Perkins-Kirkpatrick S E, Scannell H A, Sen Gupta A, Payne B L and Moore P J (2019). Marine Heatwaves Threaten Global Biodiversity and the Provision of Ecosystem Services. *Nature Climate Change*, **9** (4), 306–12.

Snelgrove P V R, Soetaert K, Solan M, Thrush S, Wei C L, Danovaro R, Fulweiler R W, Kitazato H, Ingole B, Norkko A, Parkes R J and Volkenborn N (2017). Global Carbon Cycling on a Heterogeneous Seafloor. *Trends in Ecology & Evolution*, **33** (2), 96–105.

Stirling A (2010). Keep It Complex. *Nature*, **468** (7327), 1029–31.

Strong A L, Kroeker K J, Teneva L T, Mease L A and Kelly R P (2014). Ocean Acidification 2.0: Managing Our Changing Coastal Ocean Chemistry. *BioScience*, **64** (7), 581–92.

Sukumaran S, Vijapure T, Kubal P, Mulik J, Rokade M A, Salvi S, Thomas J and Naidu V S (2016). Polychaete Community of a Marine Protected Area along the West Coast of India—Prior and Post the Tropical Cyclone, *Phyan*. *PloS One*, **11** (8), e0159368.

Temmerman S, Meire P, Bouma T J, Herman P M J, Ysebaert T and De Vriend H J (2013). Ecosystem-Based Coastal Defence in the Face of Global Change. *Nature*, **504** (7478), 79–83.

Thrush S F, Dayton P, Cattaneo-Vietti R, Chiantore M, Cummings V, Andrew N, Hawes I, Kim S, Kvitek R and Schwarz A M (2006). Broad-Scale Factors Influencing the Biodiversity of Coastal Benthic Communities of the Ross Sea. *Deep-Sea Research Part II: Topical Studies in Oceanography*, **53** (8–10), 959–71.

Thrush S F, Lewis N, Le Heron R, Fisher K T, Lundquist C J and Hewitt J (2016). Addressing Surprise and Uncertain Futures in Marine Science, Marine Governance, and Society. *Ecology and Society*, **21** (2), 44.

Tricklebank K A, Grace R V and Pilditch C A (2020). Decadal Population Dynamics of an Intertidal Bivalve (*Austrovenus stutchburyi*) Bed: Pre- and Post- a Mass Mortality Event. *New Zealand Journal of Marine and Freshwater Research*, 1–23.

Vetter E W and Dayton P K (1999). Organic Enrichment by Macrophyte Detritus, and Abundance Patterns of Megafaunal Populations in Submarine Canyons. *Marine Ecology Progress Series*, **186**, 137–48.

Waldbusser G G and Salisbury J E (2014). Ocean Acidification in the Coastal Zone from an Organism's Perspective: Multiple System Parameters, Frequency Domains, and Habitats. *Annual Review of Marine Science*, **6**, 221–47.

Wethey D S, Woodin S A, Berke S K and Dubois S F (2016). Climate Hindcasts: Exploring the Disjunct Distribution of *Diopatra biscayensis*. *Invertebrate Biology*, **135** (4), 345–56.

Wethey D S, Woodin S A, Hilbish T J, Jones S J, Lima F P and Brannock P M (2011). Response of Intertidal Populations to Climate: Effects of Extreme Events versus Long Term Change. *Journal of Experimental Marine Biology and Ecology*, **400**, 132–44.

Widdicombe S and Spicer J I (2008). Predicting the Impact of Ocean Acidification on Benthic Biodiversity: What Can Animal Physiology Tell Us? *Journal of Experimental Marine Biology and Ecology*, **366** (1–2), 187–97.

Woodin S A, Hilbish T J, Helmuth B, Jones S J and Wethey D S (2013). Climate Change, Species Distribution Models, and Physiological Performance Metrics: Predicting When Biogeographic Models Are Likely to Fail. *Ecology and Evolution*, **3** (10), 3334–46.

CHAPTER 13

Restoration of soft-sediment habitats

13.1 Introduction

One of the tragic elements of marine ecology is that we often end up documenting the decline in marine ecosystems. The biodiversity crisis and the loss of ecosystem functions that underpin many ecosystem services are important issues that need to inform our environmental futures. Similarly, informing managers about environmental change and the need to reduce stressor loads are important applications of marine science. But species are becoming functionally extinct and ecosystems are in decline.

The concern for the future of ocean ecosystems increasingly resonates among policy makers and a greater public, which emphasises the importance of reversing the negative trends. Restoration, which specifically aims to flip these negative trends into positive ones, offers the real opportunity to inject optimism into marine futures (McAfee et al., 2019). The United Nations General Assembly declared 2021–2030 the UN Decade on Ecosystem Restoration (United Nations Environment Programme, 2019). At the forefront of marine restoration has been the restoration of coral reefs (Precht, 2006), although the need for basic ecological information is still paramount (Omori, 2019). This has led to wider engagement in marine restoration and protection, along with the wider application of mitigation measures including a switch from destructive fishing practices. These activities have also led to increased tourism and engaged communities. Faced with climate change, biodiversity loss and the sustainability crisis, the management agenda is transforming from one centred on resource exploitation to the need to protect and restore.

The need to restore soft-sediment habitats draws on our understanding of ecosystem processes and the natural history of soft-sediment organisms. We can think of two basic kinds of restoration: one is a process of reducing the stress(es) on a location and waiting for the natural recovery processes to return the system to a richer state. This is the intent of marine protected areas (MPA) that seek to limit ecosystem impacts such as fishing, as well as large-scale management programmes such as Helcom (https://helcom.fi/about-us/convention/). The other process is active restoration: this is an activity designed to give nature a helping hand to overcome barriers to recovery (Figure 13.1). The roots of active restoration link back to human activities seeking to increase food resources, tracing the origins of restoration back millennia to enhancement practices such as the clam gardens of northwest America (Toniello et al., 2019). But ecological restoration is a new and evolving science (Bullock et al., 2011; Martin, 2017).

In the context of marine soft sediments, restoration to enhance ecological values is a relatively new practice that seeks to redress the impacts of habitat loss or fragmentation; declines in biodiversity, population size, viability and species ranges and loss of ecosystem goods and services. We have been slow to think about and engage in restoration in marine ecosystems compared to terrestrial environments probably because we have tended to think of our marine ecosystems as vast, well connected and too big to fail. Moreover, we have not always had sufficient data on ecological trends and a sufficient understanding of the natural history of species to ensure restoration success. The active and passive forms of restoration can work together to

Ecology of Coastal Marine Sediments: Form, Function, and Change in the Anthropocene. Simon F. Thrush, Judi E. Hewitt, Conrad A. Pilditch and Alf Norkko, Oxford University Press (2021). © Simon Thrush, Judi Hewitt, Conrad Pilditch, and Alf Norkko.
DOI: 10.1093/oso/9780198804765.003.0013

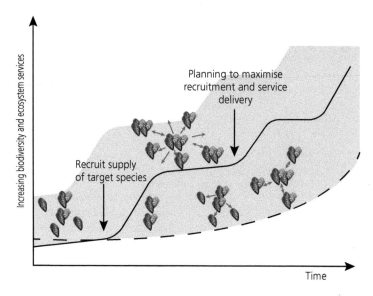

Figure 13.1 The natural recovery process after reductions in human-generated stressors can be slow, and there may be bottlenecks that need to be overcome before recovery can progress. This might include factors such as habitat suitability, recruit supply or the need for a specific combination of events. Active restoration seeks to jump-start the process by overcoming these bottlenecks; this might include the introduction of lost species or the creation of multiple habitat patches to enhance connectivity and accelerate recovery. With mussel illustrations by Dieter Tracey, IAN Image Library (ian.umces.edu/imagelibrary/).

make critical contributions to broader management strategies such as ecosystem-based management. But active restoration also plays an important part in engaging people in activities that enhance their environment and their knowledge, and that take active steps towards effective marine stewardship. This is a critical social-ecological role for restoration, helping us to navigate to more sustainable futures.

In this final chapter we are going to consider how our understanding of fundamental ecological processes underpins active restoration. This means working out the what, why, where and how of restoration. Answering these questions will draw on elements of the previous chapters, e.g. where specific animals or plants can live, how they influence each other, how they affect ecosystem function, the processes of recovery and succession, how we design and conduct studies to assess these factors and how we measure ecological restoration success. But let's not forget that effective restoration requires community, legislative, policy and management support and integration within broader manage-

ment strategies. We will touch on these broader connections within the social-ecological system at the end of the chapter.

13.2 The what, why, how and where of restoration

13.2.1 What?

Choosing species or groups of species to restore. Choice often focusses on habitat-forming species, probably due to their visibility to those living near and working in and around soft-sediment systems. The restoration of marine species initially followed terrestrial ecosystems which has focussed on restoring ecosystems through the planting of trees and shrubs. Thus, much of the early marine efforts have focussed on the fringing vegetative habitats of salt marsh and mangrove or seagrasses. The soft-sediment animals that have been the target of restoration are reef-forming shellfish, predominantly oysters and to a lesser extent mussels—highly visible and utilised species.

It would be easy to assume that we restore what we have lost, but there are usually complicating factors to consider. We may not have sufficient knowledge of the species biology to create successful opportunities and changes in habitats may mean that the species can no longer survive at the location from which it was lost. Future environmental changes also need considering, especially changes associated with climate change. Sometimes, the native species are no longer a viable option, but the use of substitute species can be very contentious.

13.2.2 Why?

Single species to biodiversity and ecosystem service benefits. Being clear on the multiple ecological benefits that can be derived from restoration is often important to help garner support and it is vital to the project design and the metrics used to define success. Ecosystem services are recognised as a critical factor defining why we need to restore marine species and habitats (Biggs et al., 2012). Many soft-sediment organisms modify habitats affecting biogeochemical processes, providing structure and affecting boundary flows (see Chapter 2). In some cases, specific services are targeted for restoration, for example the role that filter-feeding shellfish or seagrass beds can play in affecting water clarity. This means we need to develop techniques that allow us to scale up ecosystem service delivery from *in situ* measurements to deliver estimates of broader-scale effects. Processes that lead to the delivery of ecosystem services often do not necessarily scale up on a per-capita basis, because of interactions across multiple scales (Hewitt et al., 2007). For example, one shellfish filtering at 1 l/hour does not mean that 1,000 shellfish will filter 1 m³/hour, nor do their effects on water clarity necessarily scale directly. The filtration rate and the potential for deposition or resuspension of sediment might be affected by boundary flow velocities, the quality of food resources that the mussels can extract from the water and the turbulence generated by flow–structure interactions between shells, clumps and patches. Although broad-scale models can provide valuable insight into lost functionality in ecosystems (zu Ermgassen et al., 2013), linking processes

across scale is important to develop models that can forecast restoration success.

The 'why' question can also include whether the restoration is focussed on a single habitat or location, or whether restoration is hoped to spread to the larger ecosystem. Often the benefits of restoration are not only manifest in the restored habitat but more widely. For example, coasts and estuaries often provide important nursery habitats for fish and many other species and are warm productive areas providing good conditions for growth. Species that form habitat on the seafloor such as shellfish reefs, seagrass or sponge gardens can provide refugia from predators and adverse environmental conditions (Thrush & Dayton, 2010). Thus, restoring these nursery habitats will not only generate local hotspots of productivity and diversity but also potentially enhance populations over much wider scales. A review of fisheries of the North Atlantic (Seitz et al., 2014) identified that 44% of the exploited species, as considered by ICES, use coastal habitats for at least part of their life cycle. These coastal habitats contribute important economic benefits to fisheries, contributing 77% of the commercial landings of ICES.

Other demonstrations of the potential for restoration 'spread' include a combination of hydrodynamic modelling of larval dispersal and genetic parentage analysis that allowed Le Port et al. (2017) to demonstrate 10.6% of newly settled juvenile snapper sampled up to 40 km outside of an MPA were the offspring of adults living within the 5.2 km² no-take MPA. In another study, published data reporting fish and mobile invertebrate abundance on an oyster reef and in an unstructured control habitat on the Atlantic and Gulf coasts of the USA were used to derive estimates of abundance and productivity (zu Ermgassen et al., 2016). This analysis highlighted that the majority of production could be attributed to the reef and that in just two to three years, the productivity of fish and mobile invertebrates was almost doubled in both regions (Figure 13.2). However, it is important to note that the scale of restoration spread may change over the time of the restoration effort. Analysis by zu Ermgassen et al. (2016) assumes that juvenile habitat is limiting and this (hopefully) will change with restoration success.

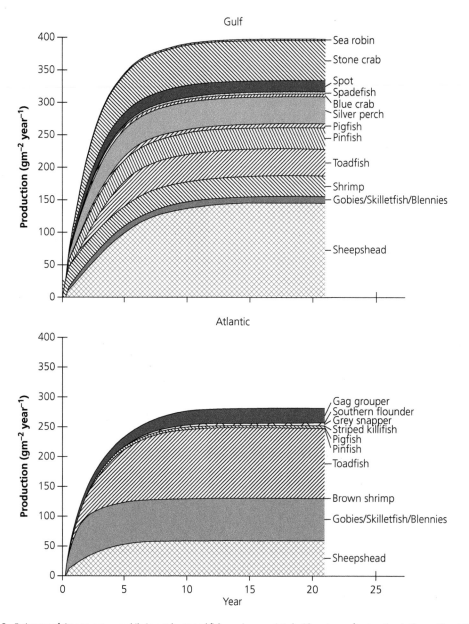

Figure 13.2 Estimates of time to restore mobile invertebrate and fish species associated with oyster reef restoration in the southeast USA. From zu Ermgassen et al. (2016). Reprinted under Creative Commons Attribution 4.0 International (CC BY 4.0) licence.

13.2.3 How?

Requires we address the potential of different life stages, the size and spatial arrangement of patches of organisms outplanted and the scales needed to ensure long-term viability and potentially expansion of restored species and habitats. It may be possible to obtain individuals from a source population, but it is important to ensure that we do not simply move the restoration need from one place to another. Organisms for restoration may also be sourced from nurseries, hatcheries or aquaculture, but these facilities themselves may need to be developed if the target species is not of commercial interest.

Depending on the species, different life stages may be more amenable to restoration. Transplanting procedures might target larval, juvenile or adult life stages. Larvae may be easy to source, but larvae at a stage that is competent to settle to the sediment are likely to have specific habitat requirements and natural mortality rates of these early life stages are typically very high. Juveniles' life stages, perhaps associated with a post-larval dispersal stage, may be more robust, but will still suffer high mortality rates and require more time to grow, reproduce and deliver ecosystem function. Adults may require multiple years before they are ready for use, and be expensive to transport, but are likely robust to both transportation and outplanting and once established should contribute to ecosystem functions and start to reproduce.

Once the species and life stage are selected, we need to develop outplanting techniques. Here disturbance, succession and landscape ecology can help us mimic natural spatial patterns, identify techniques to facilitate establishment, predict time scales of restoration success and identify ecological processes such as natural disturbance or predation that might limit success. Ideally restored patches will contribute to the population dynamics either through self-recruitment or by a contribution to the meta-population. Giving nature a hand to get started and then relying on natural processes to expand the extent of restoration will be essential in meeting restoration targets for large areas of the seafloor.

13.2.4 Where?

Defining habitat requirements, maximising recruitment success, delivering specific ecosystem service benefits. Whatever life stage is selected for restoration we will need to consider its habitat requirements along with any subsequent life stages. Species distribution models (Dormann et al., 2012) can help identify suitable sites, if relevant biological and environmental data are available. This might include information on hydrodynamics, sediment type, suspended sediment concentrations, environmental stressors, nutrients or food resources.

Interactions with other species in the location may also be important. For example, predation is a major source of mortality for shellfish, especially as habitat disturbance and other human activities reduce refugia and increase the density of predators/scavengers on the seafloor (Talman et al., 2003; Thrush & Dayton, 2010). Both prey density and the stress levels of transplanted species are likely to affect predation rates (Hines et al., 1997, 2009).

Decisions based on meeting the best environmental and biological site selection criteria may result in locating restoration sites away from historic locations of the restored species. But often the location for restoration is chosen by local communities and present habitats might not be now ideal. In such cases, it may be possible to use localised habitat modifications to enhance restoration success, for example, providing settlement surfaces, stabilising sediments, elevating organisms to limit impacts from suspended sediment and creating protection from predation. Ideally site selection criteria will be informed by data collected on natural beds of the restored species, but this may not always be possible and we will then need to rely on other lines of evidence to help improve the success of our restoration efforts.

Ultimately restoration efforts should seek to restore ecosystem functions, which means restoring ecosystem interaction networks across multiple scales. This includes restoring the full complement of traits within communities and physicochemical conditions to ensure the performance of multiple functions within patches. Connectivity across landscapes needs to be restored to ensure resilience of communities, populations and functions, and, in particular, the capacity of natural processes to continue to enhance the recovery trajectories and long-term viability. To ensure restoration of coastal areas at these broader space and time scales, coordination in managing stressors associated with land- and sea-based activities is required.

13.3 Restoration, bioremediation and ecological infrastructure

The what, why, where and how of restoration are not simple questions and are tightly coupled. For example, experiments in restoring oyster reefs in the Yangtze estuary demonstrated that the biodiversity benefits (why) increased with salinity (where) (Liu et al., 2018). The definition of restoration goals is critical for active restoration, but these goals should be as widely considered as possible. Even when the voiced desire is to restore a specific species or set of services (e.g. enhanced fisheries), co-benefits will often accrue. For example, restoring shellfish reefs may enhance nursery habitat for fish but also affect sediment biogeochemistry, thus linking to services associated with productivity, biodiversity, water clarity and mitigating eutrophication. In some cases, the end goal may require restoration of most ecosystem functionality. In seeking to enhance the delivery of many services or ecosystem functionality, we might need to consider restoration of infaunal species as well as the habitat formers such as mangroves, seagrass, oysters and mussels.

These justifications for restoration, and the need to modify existing environmental conditions, provide a link to bioremediation. Bioremediation involves directly or indirectly enhancing organisms in the ecosystem for their specific role in contaminant breakdown or removal from reactive environments (Gonzalez et al., 2019; Shen et al., 2016, 2017) (also see Chapter 11). There is sometimes the temptation to involve non-native species, especially when they prove able to cope with more degraded environments than the native species (see Chapter 11); however, this requires careful consideration as the introduction of non-native species frequently has negative unintended consequences.

There are also links to ecological infrastructure. Fringing habitats have been demonstrated to baffle wave energy and thus offer natural processes as solutions to reduce the impacts of storm surge and sea-level rise. Further offshore shellfish reefs and large seaweeds can also dampen wave energy. These species offer alternatives to grey-infrastructure coastal defences and have been termed 'working with nature'. These approaches are not relevant in all locations as the requisite species' habitat requirements may not suit or the potential of the natural features to absorb wave or tide energy may not be sufficient. Nevertheless, these are sustainable solutions often with more co-benefits and better cost–benefit

ratios than conventional coastal engineering (Solan et al., 2020). Consequently, they are now being used around the world and in some situations at scale (Temmerman et al., 2013; Van Coppenolle & Temmerman, 2020) (see also Chapter 12).

13.4 Restoration in ecosystem networks

The ecosystems within which we engage in restoration actions are driven by networks of interactions and feedback processes (Suding et al., 2004; Suding & Hobbs, 2009). These interactions sometimes result in hysteresis, meaning that the time scales of recovery can be slower than those of habitat loss and disturbance. Thus, our ability to identify networks of drivers of ecosystem function informs the strategy for restoration (Nyström et al., 2012), in particular whether 'turning off the tap' will allow for improvement over time scales desired by society, or whether the ecosystem needs a helping hand. Active restoration of soft-sediment habitats is still very much an emerging practice, but for seagrass, salt marsh and shellfish, positive interactions and feedback processes are important for success (Grabowski et al., 2005; Silliman et al., 2015; Suykerbuyk et al., 2012). This is because of the importance of facilitation in soft sediments and the role positive species interactions can play under stressful environmental conditions. However, a survey of restoration organisations using plants such as salt marsh, mangrove and seagrass found that almost all of these organisations disperse propagules to reduce competition between individual plants (Silliman et al., 2015). Understanding natural recruitment dynamics and the spatial structure of populations of the target species can provide useful information on long-term recovery.

The mechanisms of facilitation not only depend on initial organism densities but may also be influenced by spatial patterning (van de Koppel et al., 2005). The spatial scale of restoration is also important; it may be more effective to disperse outplanted species into clumps that match the multi-scale fractal patterns of natural beds to maximise density over a smaller area. This may allow for local facilitation while not enhancing densities in ways that support high predator density (Hewitt & Cummings, 2013).

Many species are unable to move once established as they root themselves into the sediment or glue themselves into clumps, but others are able to move and even a low degree of mobility can allow animals to self-organise into clumps. Adult mussels will do this to quickly generate spatial patterns similar to those seen in natural mussel beds (e.g. Commito et al., 2016). A combination of field observations, modelling and flume experiments allowed Bertolini et al. (2019) to identify two mechanisms working at different scales that generated local clumping. This local clumping reduced the potential for hydrodynamic energy to remove individuals and increased turbulence between patches which in turn increased food supply by mixing water layers. These mechanisms may drive restoration success at study sites but under a different suite of conditions other mechanisms may be more relevant. Understanding the mix of environmental conditions that contribute to mechanistic processes, and the scale at which they do this (see Chapters 1, 4 and 7), is crucial to developing successful restoration!

The functional extinction of species can lead to a tipping point or regime shift; if this happens, it is clear evidence that the lost species were important influencers of ecosystem dynamics. This means there are good reasons to want to restore habitats. However, these regime shifts can leave legacy effects that need to be addressed if restoration is to be successful. For example, many shellfish and other epifaunal species that establish on the seafloor require a settlement site that lifts the recruits off the seabed—this can reduce the chance of predation or smothering by fine sediments. Bivalve shells often provide these small hard surfaces for attachment. If the shells are lost or smothered with sediment this may limit recruitment. Similarly, seagrasses often act as traps for fine sediment. If the seagrass is lost this sediment may be constantly being deposited and eroded, elevating the turbidity of the water column and making it harder for seedlings to attach and plants to photosynthesise (Van der Heide et al., 2007). These hysteresis effects will depend on the target species, its natural history and the context dependency of specific sites' environmental characteristics.

Box 13.1 Measuring success

Active restoration of soft-sediment habitats is in its infancy and thus measuring the success of restoration is a critical strategy of adaptive development and implementation. Some restoration trials may fail but we can learn from these to develop new methods (de Paoli et al., 2015). We need to consider 'success' as a series of steps rather than an end point. These steps can include the growth and viability of the transplanted organisms, recruitment into the habitat or adjacent ones, enhancement of biodiversity and ecosystem services (Basconi et al., 2020) and, finally, a dynamic, resilient, fully functional ecosystem.

Measuring success will also depend on space and time scales (zu Ermgassen et al., 2012)—as the scale of our restoration aspiration increases so will the extent and complexity of monitoring. Shifting baselines (Duarte et al., 2009; see also Chapter 11) can mean that we will need to move restoration efforts to locations that increase the chances of success. For example, if disturbance has homogenised benthic communities there may be a limited natural supply of the late successional stages species that are likely to be the

targets of restoration. Even if some of the species can be provided by off-site sources, the full range of species may be unable to recover due to supply-dispersal limitations. Similarly, historical or predicted future habitat change may also influence the suitability of sites and species. We may simply not have the background data on how a species we are trying to restore used to live and what role it played in the ecosystem (Gillies et al., 2017). Ideally these issues will be factored into the restoration project design, but they may also need to be incorporated into success measures because removing human pressures to return to more pristine conditions may not always be possible. Disturbance recovery dynamics can also provide some insight into the time scales we should expect to consider in terms of restoration. Reviewing 51 long-term projects, Borja et al. (2010) identified recovery time scales of 10–25 years. This might be relevant for short-lived and short-succession processes, but as the recovery becomes more complex and the age of the restored species increases, we can expect these time scales to increase.

13.5 Linking to social engagement and blue economies

Restoration can enhance biodiversity and ecosystem services but it is also an important cultural, social and economic activity (Box 13.1). Restoring culturally treasured species can be an act of cultural restoration as much as an ecological action (Dam Lam et al., 2019; Mathews & Turner, 2017; Paul-Burke et al., 2018; Thurstan et al., 2020). More broadly communities often want to engage in actions to improve their environment. We have seen these developments enhance engagement with marine environmental issues and management. Three US restoration case studies that differed in the degree of state or public initiation helped to identify the factors that sustain public support for restoration, revealing the importance of public demand, political action and critical support of local stakeholders along with ecologically credible planning and scientific monitoring of progress (DeAngelis et al., 2020). Similarly, considering existing bio-physical and social-economic conditions of the site, and having an area-specific restoration and community development plan with active participation of the local community along with having the right ecological knowledge were essential in the restoration of mangroves in South-East Asia (Biswas et al., 2009).

Some of the ecosystem benefits of restoration (e.g. nursery habitat, enhanced recruit supply to exploited populations, shoreline protection) can directly translate into monetary benefits and generate new restoration job markets and economies. For example, in Half Moon Reef, Texas, the restoration of 54 acres of shellfish created 12 new jobs, approximately $US 0.5 M in annual labour income, with $US 1.3 M in total value added to the economy from enhanced recreational fishing alone (Carlton et al., 2016). The economic value of oyster reefs in the USA, excluding economic value from harvesting, has been conservatively estimated at $US 5,500–99,000 per ha per year (Grabowski et al., 2012). Importantly these ecosystem service economic values were not realised if the oyster reefs were destructively harvested. Habitat restoration projects in the USA generate, on average, 17 jobs per million dollars spent, which is similar to job generation by parks and land conservation, and higher than grey infrastructure such as coal, gas and nuclear energy generation. Broader-scale economic benefits associated with fisheries, coastal tourism and property values and better water quality amplify the benefits to society (Edwards et al., 2013). In Australia, it has been estimated that the repair of Australia's shellfish reefs, seagrasses and salt marshes would cost about $AU 350 M, a small sum in comparison to upgrading grey infrastructure projects such as roads or rail, with the return on investment reached in five years, based on improvements to fisheries alone (Gillies et al., 2015).

13.6 Close out

We have raised many questions that will need to be addressed if we are to successfully restore the range of degraded soft-sediment habitats. This represents an opportunity and a challenge to apply ecological knowledge and processes. Recognising that ecology alone will not drive successful restoration highlights the importance of interdisciplinary and transdisciplinary research to successfully implement this new restoration paradigm (Bersoza Hernández et al., 2018). This will require new models weaving together multiple knowledge systems with new understanding emerging from the interface (Figure 13.3).

Although we have focussed on 'active' restoration, any restoration needs to consider whether such active restoration is needed and will be successful. Regardless, reduction in stressors will generally be required before active efforts are likely to be successful. Furthermore, as the spatial scale of the degradation increases in size, so too does the likelihood of reduction in stressors being the primary mechanism for restoration. At the same time, connectivity to undegraded areas will be decreased and so will the potential for natural recolonisation. Nested 'active' efforts within large-scale management focussed on reduction of stressors may be useful. Research and models focussing on the scaling up of restoration within a landscape of degradation are an interesting area of research.

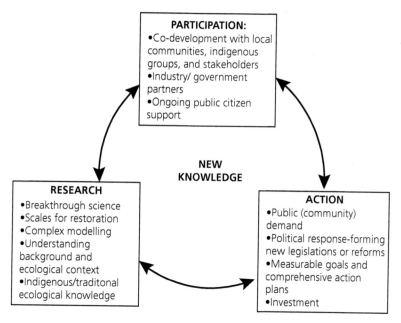

Figure 13.3 New models that create active interdependences between research and society and link to effective action generate new knowledge to restore coastal ecosystems.

References

Basconi L, Cadier C and Guerrero-Limón G (2020). Challenges in Marine Restoration Ecology: How Techniques, Assessment Metrics, and Ecosystem Valuation Can Lead to Improved Restoration Success. In: Jungblut S, Liebich V and Bode-Dalby M, eds. *YOUMARES 9 - the Oceans: Our Research, Our Future: Proceedings of the 2018 Conference for YOUng MArine RESearcher in Oldenburg, Germany,* Springer International, Cham, pp. 83–99.

Bersoza Hernández A, Brumbaugh R D, Frederick P, Grizzle R, Luckenbach M W, Peterson C H and Angelini C (2018). Restoring the Eastern Oyster: How Much Progress Has Been Made in 53 Years? *Frontiers in Ecology and the Environment,* **16** (8), 463–71.

Bertolini C, Cornelissen B, Capelle J, van de Koppel J and Bouma T J (2019). Putting Self-Organization to the Test: Labyrinthine Patterns as Optimal Solution for Persistence. *Oikos,* **128** (12), 1805–15.

Biggs R, Schlüter M, Biggs D, Bohensky E L, Burnsilver S, Cundill G, Dakos V, Daw T M, Evans L S, Kotschy K, Leitch A M, Meek C, Quinlan A, Raudsepp-Hearne C, Robards M D, Schoon M L, Schultz L and West P C (2012). Toward Principles for Enhancing the Resilience of Ecosystem Services. *Annual Review of Environment and Resources,* **37**, 421–48.

Biswas S R, Mallik A U, Choudhury J K and Nishat A (2009). A Unified Framework for the Restoration of Southeast Asian Mangroves—Bridging Ecology, Society and Economics. *Wetlands Ecology and Management,* **17** (4), 365–83.

Borja A, Dauer D M, Elliott M and Simenstad C A (2010). Medium- and Long-Term Recovery of Estuarine and Coastal Ecosystems: Patterns, Rates and Restoration Effectiveness. *Estuaries and Coasts,* **33** (6), 1249–60.

Bullock J M, Aronson J, Newton A C, Pywell R F and Rey-Benayas J M (2011). Restoration of Ecosystem Services and Biodiversity: Conflicts and Opportunities. *Trends in Ecology and Evolution,* **26** (10), 541–9.

Carlton J S, Ropicki A and Balboa B (2016). The Half Moon Reef Restoration: A Socioeconomic Evaluation. Sea Grant Texas, *TNC Report, Texas A&M University.*

Commito J A, Gownaris N J, Haulsee D E, Coleman S E and Beal B F (2016). Separation Anxiety: Mussels Self-Organize into Similar Power-Law Clusters Regardless of Predation Threat Cues. *Marine Ecology Progress Series,* **547**, 107–19.

Dam Lam R, Gasparatos A, Chakraborty S, Rivera H and Stanley T (2019). Multiple Values and Knowledge Integration in Indigenous Coastal and Marine Social-Ecological Systems Research: A Systematic Review. *Ecosystem Services,* **37**, 100910.

de Paoli H, van de Koppel J, van der Zee E, Kangeri A, van Belzen J, Holthuijsen S, van den Berg A, Herman P, Olff H and van der Heide T (2015). Processes Limiting Mussel Bed Restoration in the Wadden-Sea. *Journal of Sea Research*, **103**, 42–9.

DeAngelis B M, Sutton-Grier A E, Colden A, Arkema K K, Baillie C J, Bennett R O, Benoit J, Blitch S, Chatwin A, Dausman A, Gittman R K, Greening H S, Henkel J R, Houge R, Howard R, Hughes A R, Lowe J, Scyphers S B, Sherwood E T, Westby S and Grabowski J H (2020). Social Factors Key to Landscape-Scale Coastal Restoration: Lessons Learned from Three U.S. Case Studies. *Sustainability (Switzerland)*, **12** (3), 869.

Dormann C F, Schymanski S J, Cabral J, Chuine I, Graham C, Hartig F, Kearney M, Morin X, Römermann C, Schröder B and Singer A (2012). Correlation and Process in Species Distribution Models: Bridging a Dichotomy. *Journal of Biogeography*, **39** (12), 2119–31.

Duarte C M, Conley D, Carstensen J and Sanchez-Comacho M (2009). Return to Neverland: Shifting Baselines Affect Eutrophication Restoration Targets. *Estuaries and Coasts*, **32**, 29–36.

Edwards P E T, Sutton-Grier A E and Coyle G E (2013). Investing in Nature: Restoring Coastal Habitat Blue Infrastructure and Green Job Creation. *Marine Policy*, **38**, 65–71.

Gillies C L, Crawford C and Hancock B (2017). Restoring Angasi Oyster Reefs: What Is the Endpoint Ecosystem We Are Aiming for and How Do We Get There? *Ecological Management and Restoration*, **18** (3), 214–22.

Gillies C L, Creighton C and McLeod I M (2015). Shellfish Reef Habitats: A Synopsis to Underpin the Repair and Conservation of Australia's Environmentally, Socially and Economically Important Bays and Estuaries. Report to the National Environmental Science Programme, Marine Biodiversity Hub. *Centre for Tropical Water and Aquatic Ecosystem Research (TropWATER) Publication*, James Cook University, Townsville.

Gonzalez S V, Johnston E, Gribben P E and Dafforn K (2019). The Application of Bioturbators for Aquatic Bioremediation: Review and Meta-Analysis. *Environmental Pollution*, **250**, 426–36.

Grabowski J H, Brumbaugh R D, Conrad R F, Keeler A G, Opaluch J J, Peterson C H, Piehler M F, Powers S P and Smyth A R (2012). Economic Valuation of Ecosystem Services Provided by Oyster Reefs. *BioScience*, **62** (10), 900–9.

Grabowski J H, Hughes A R, Kimbro D L and Dolan M A (2005). How Habitat Setting Influences Restored Oyster Reef Communities. *Ecology*, **86** (7), 1926–35.

Hewitt J E and Cummings V J (2013). Context-Dependent Success of Restoration of a Key Species, Biodiversity and Community Composition. *Marine Ecology Progress Series*, **479**, 63–73.

Hewitt J E, Thrush S F, Dayton P K and Bonsdorf E (2007). The Effect of Scale on Empirical Studies of Ecology. *The American Naturalist*, **169**, 398–408.

Hines A H, Long C W, Terwin J R and Thrush S F (2009). Facilitation, Interference, and Scale: The Spatial Distribution of Prey Patches Affects Predation Rates in an Estuarine Benthic Community. *Marine Ecology Progress Series*, **385**, 127–35.

Hines A H, Whitlatch R B, Thrush S F, Hewitt J E, Cummings V J, Dayton P K and Legendre P (1997). Nonlinear Foraging Response of a Large Marine Predator to Benthic Prey: Eagle Ray Pits and Bivalves in a New Zealand Sandflat. *Journal of Experimental Marine Biology and Ecology*, **216**, 211–28.

Le Port A, Montgomery J C, Smith A N H, Croucher A E, McLeod I M and Lavery S D (2017). Temperate Marine Protected Area Provides Recruitment Subsidies to Local Fisheries. *Proceedings of the Royal Society B: Biological Sciences*, **284** (1865), 2017.1300.

Liu Z, Yu P, Chen M, Cai M, Fan B, Lv W, Huang Y, Li Y and Zhao Y (2018). Macrobenthic Community Characteristics and Ecological Health of a Constructed Intertidal Oyster Reef in the Yangtze Estuary, China. *Marine Pollution Bulletin*, **135**, 95–104.

McAfee D, Doubleday Z A, Geiger N and Connell S D (2019). Everyone Loves a Success Story: Optimism Inspires Conservation Engagement. *BioScience*, **69** (4), 274–81.

Martin D M (2017). Ecological Restoration Should Be Redefined for the Twenty-First Century. *Restoration Ecology*, **25** (5), 668–73.

Mathews D L and Turner N J (2017). Ocean Cultures: Northwest Coast Ecosystems and Indigenous Management Systems. In: Levin P S and Poe M R, eds. *Conservation for the Anthropocene Ocean: Interdisciplinary Science in Support of Nature and People*, Academic Press, Cambridge, MA, pp. 169–206.

Nyström M, Norström A, Blenckner T, de la Torre-Castro M, Eklöf J, Folke C, Österblom H, Steneck R, Thyresson M and Troell M (2012). Confronting Feedbacks of Degraded Marine Ecosystems. *Ecosystems*, **15** (5), 695–710.

Omori M (2019). Coral Restoration Research and Technical Developments: What We Have Learned So Far. *Marine Biology Research*, **15** (7), 377–409.

Paul-Burke K, Burke J, Bluett C and Senior T (2018). Using Māori Knowledge to Assist Understandings and Management of Shellfish Populations in Ōhiwa Harbour, Aotearoa New Zealand. *New Zealand Journal of Marine and Freshwater Research*, **52** (4), 542–56.

Precht W F, ed. (2006). *Coral Reef Restoration Handbook*, CRC Press, Boca Raton, FL.

Seitz R D, Wennhage H, Bergström U, Lipcius R N and Ysebaert T (2014). Ecological Value of Coastal Habitats for Commercially and Ecologically Important Species. *ICES Journal of Marine Science*, **71** (3), 648–65.

Shen H, Jiang G, Wan X, Li H, Qiao Y, Thrush S and He P (2017). Response of the Microbial Community to Bioturbation by Benthic Macrofauna on Intertidal Flats. *Journal of Experimental Marine Biology and Ecology*, **488**, 44–51.

Shen H, Thrush S F, Wan X, Li H, Qiao Y, Jiang G, Sun R, Wang L and He P (2016). Optimization of Hard Clams, Polychaetes, Physical Disturbance and Denitrifying Bacteria of Removing Nutrients in Marine Sediment. *Marine Pollution Bulletin*, **110** (1), 86–92.

Silliman B R, Schrack E, He Q, Cope R, Santoni A, Van Der Heide T, Jacobi R, Jacobi M and Van De Koppel J (2015). Facilitation Shifts Paradigms and Can Amplify Coastal Restoration Efforts. *Proceedings of the National Academy of Sciences of the United States of America*, **112** (46), 14295–300.

Solan M, Bennett E M, Mumby P J, Leyland J and Godbold J A (2020). Benthic-Based Contributions to Climate Change Mitigation and Adaptation. *Philosophical Transactions of the Royal Society B: Biological Sciences*, **375** (1794), 2019.0107.

Suding K N, Gross K L and Houseman G R (2004). Alternative States and Positive Feedbacks in Restoration Ecology. *Trends in Ecology & Evolution*, **19** (1), 46–53.

Suding K N and Hobbs R J (2009). Threshold Models in Restoration and Conservation: A Developing Framework. *Trends in Ecology & Evolution*, **24**, 271–9.

Suykerbuyk W, Bouma T J, Van Der Heide T, Faust C, Govers L L, Giesen W B J T, De Jong D J and Van Katwijk M M (2012). Suppressing Antagonistic Bioengineering Feedbacks Doubles Restoration Success. *Ecological Applications*, **22** (4), 1224–31.

Talman S G, Norkko A, Thrush S F and Hewitt J E (2003). Habitat Structure and the Survival of Juvenile Scallops (*Pecten Novaezelandiae*): Comparing Predation in Habitats with Varying Complexity. *Marine Ecology Progress Series*, **269**, 209–21.

Temmerman S, Meire P, Bouma T J, Herman P M J, Ysebaert T and De Vriend H J (2013). Ecosystem-Based Coastal Defence in the Face of Global Change. *Nature*, **504** (7478), 79–83.

Thrush S F and Dayton P K (2010). What Can Ecology Contribute to Ecosystem-Based Management? *Annual Review of Marine Science*, **2**, 419–41.

Thurstan R H, Diggles B K, Gillies C L, Strong M K, Kerkhove R, Buckley S M, King R A, Smythe V,

Heller-Wagner G, Weeks R, Palin F and McLeod I (2020). Charting Two Centuries of Transformation in a Coastal Social-Ecological System: A Mixed Methods Approach. *Global Environmental Change*, **61**, 102058.

Toniello G, Lepofsky D, Lertzman-Lepofsky G, Salomon A K and Rowell K (2019). 11,500 Y of Human–Clam Relationships Provide Long-Term Context for Intertidal Management in the Salish Sea, British Columbia. *Proceedings of the National Academy of Sciences of the United States of America*, **116** (44), 22106–14.

United Nations Environment Programme (2019). New UN Decade on Ecosystem Restoration Offers Unparalleled Opportunity for Job Creation, Food Security and Addressing Climate Change. 1 March. https://www.unenvironment.org/news-and-stories/press-release/new-un-decade-ecosystem-restoration-offers-unparalleled-opportunity.

Van Coppenolle R and Temmerman S (2020). Identifying Global Hotspots Where Coastal Wetland Conservation Can Contribute to Nature-Based Mitigation of Coastal Flood Risks. *Global and Planetary Change*, **187**, 103125.

van de Koppel J, Rietkerk M, Dankers N and Herman P M J (2005). Scale-Dependent Feedback and Regular Spatial Patterns in Young Mussel Beds. *The American Naturalist*, **165** (3), E66–77.

Van der Heide T, van Nes E H, Geerling G W, Smolders A J P, Bouma T J and van Katwijk M M (2007). Positive Feedbacks in Seagrass Ecosystems: Implications for Success in Conservation and Restoration. *Ecosystems*, **10**, 1311–22.

zu Ermgassen P S, Grabowski J H, Gair J R and Powers S P (2016). Corrigendum to: Quantifying Fish and Mobile Invertebrate Production from a Threatened Nursery Habitat, (Journal of Applied Ecology, 53, 596–606). *Journal of Applied Ecology*, **55** (6), 3005–9.

zu Ermgassen P S E, Spalding M D, Blake B, Coen L D, Dumbauld B, Geiger S, Grabowski J H, Grizzle R, Luckenbach M, McGraw K, Rodney W, Ruesink J L, Powers S P and Brumbaugh R (2012). Historical Ecology with Real Numbers: Past and Present Extent and Biomass of an Imperilled Estuarine Habitat. *Proceedings of the Royal Society B: Biological Sciences*, **279** (1742), 3393–400.

zu Ermgassen P S E, Spalding M D, Grizzle R E and Brumbaugh R D (2013). Quantifying the Loss of a Marine Ecosystem Service: Filtration by the Eastern Oyster in US Estuaries. *Estuaries and Coasts*, **36** (1), 36–43.

Glossary

Additive effects The effect of two or more stressors is the addition of each single effect

Alpha or point diversity The average or mean diversity in a habitat or area

Antagonistic effects The effect of two or more stressors is less than the effect of each singly

Anthropocene The most recent geologic time period, in which human activities have had a dominant impact on earth systems

BEF Studies on biodiversity–ecosystem function relationships

Beta diversity The ratio (or difference) between gamma and alpha diversity or the turnover

Beta diversity—nestedness The species observed at one site are a subset of species found at others

Bioadvection Transport of solutes or energy (heat) within the sediment caused by the presence of organisms

Bioirrigation Pumping of water through sediment by organisms—often used to describe the pumping of water through burrows by crustaceans and polychaetes

Biological trait Morphological or life history characteristic of an individual or species

Carbon sequestration The process of capture and long-term storage of atmospheric carbon

Chronic stressor(s) A stressor or stressors, often of low magnitude at any one time, that operates over a long period of time more or less continuously

Community A level of biological organisation involving multiple species

Complementarity (as a process) Similar to niche differentiation, species divide up a niche by processing resources more effectively

Context dependency Context dependency and generality are two sides of the same coin. We seek to find locations, environments and times over which we will observe the same responses to the same predictors

Continental shelf The extended perimeter of the continents and associated coastal plain extending to 200 metres depth

Co-variables These are variables beyond those of primary interest that are measured because they may also influence how the response variable actually responds. These variables can be used to determine the contexts in which a similar response could be expected to be observed

Cross-scale Responses at one scale can be driven by processes that interact across multiple scales

Cumulative effects Response to accumulating stressors in space and time. Can refer to one or multiple stressors

Denitrification The microbial reduction of NO_3^- to N_2, O, and of N_2O to N_2

Deposition Process by which sediment or other particles are deposited on the seafloor

Diffuse stressor A stressor that influences a large area, often over a long time. Opposite of point source stressor

Disturbance Any event that disturbs part or all of an ecosystem. It can operate at a range of organisation levels and scales and can be natural (e.g. waves) or related to human activities when the term 'stressor' is often used

Dynamic state Ecosystems and their components vary over time

Ecosystem engineers Organisms or structures produced by organisms that provide habitat for other organisms by altering the surrounding environment

Ecosystem interaction networks Communities and ecosystems consist of a number of connecting processes and components. Usually networks focus on certain subsets of the whole, e.g. a foodweb

Ecosystem service The benefits people obtain from ecosystems

Effect traits Traits that are important for ecosystem functioning

Emergent properties A property that arises at one scale as a consequence of interactions among entities at a lower level of organisation; it is a property unique to the higher level of organisation and usually not predictable from knowledge of properties at lower levels

Environmental filtering Abiotic conditions determine whether a species will persist at a site

Facilitation Describes a species interaction that benefits at least one participant and causes harm to neither

Fast and slow processes Fast processes show change over short time scales, while slow processes result in change over much longer time scales. Their definition is relative to the scale of the study question

Functional diversity Diversity of biological or functional traits

Functional extinction A decline in a functional group or trait to the point that it no longer plays a significant role

Functional group Group of biological traits that combine to create a group that would affect a specific ecosystem function

Functional redundancy Multiple species share similar roles in particular ecological functions, suggesting that if one species is impacted the function can still occur

Gamma diversity The total diversity in a landscape

Habitat structuring species Species that produce habitat structure—see also ecological engineers

Heterogeneity Variability in spatial or temporal arrangement

Hierarchical scale Processes are often considered to act in a hierarchy. Environmental processes (e.g. temperature, depth) operate at the broad scale, setting the scene for medium population-based processes (e.g. recruitment), within which are nested individual organism processes (e.g. feeding)

Hypoxia Low levels of oxygen, but not yet anoxic

Hysteresis The pathway to recovery differs from the degradation path; often including a lag effect

Insurance hypothesis or effect Having more species creates a greater likelihood that some of them will maintain ecosystem functioning even if others fail

Key species A high-performing species in a community which affects the overall biodiversity, community structure or ecosystem functioning

Meta-communities A set of interacting communities which are linked by the dispersal of multiple, potentially interacting species

Microphytobenthos Microscopic algae living on or in the sediment

Multifunctionality Ecosystem functions are generally studied separately but increasingly there are attempts to consider multiple functions at the same time

Multiplicative stressors As opposed to additive, covering both synergistic and antagonistic

Multi-scale Many so-called small processes can create effects at larger scales, e.g. feeding currents and siphons interact with flow to create benthic boundary dynamics that can extend from mm to km, affecting recruitment success of larvae and elevating organic content of the sediment

Multi-year cycles Across-year changes in seasonal patterns. These can occur on the order of 3–15 years or even longer, depending on the drivers

Niche partitioning The process by which natural selection drives competing species into different patterns of resource use or different niches

Permeability The resistance to water flow through the sediment

Perturbation (a response) A temporal change in ecosystem, community or population structure or resources as a result of disturbance or other environmental or ecological forcing

Phenotypic plasticity The ability of one genotype to produce more than one phenotype when exposed to different environments

Physiological tolerances Range of environmental conditions within which a species can feed, reproduce and grow

Plasticity The ability of an organism to exhibit different behaviours in different environments

Porewater Water contained in spaces between sediment particles

Porosity The amount of empty spaces between particles in the sediment

Pseudo-replication Replicates that are not independent. This is a tricky point as replicates need to be sufficiently similar that they are actually replicates but not located so close in space or time that one replicate affects what is observed in another replicate

Rank abundance Abundance of species in a study is ranked in order of abundance and plotted

Rarity Many species are rare, that is, they occur in low numbers, occupy a restricted range of sites or occur infrequently in temporal sampling. Often subdivided into uniques (occurring once), duplicates (occurring twice), singletons (only one individual), doubletons (only two individuals)

Recovery (in the context of disturbance) The colonisation and growth of organisms in a disturbed patch. The population, community or ecosystem function in the patch may be considered recovered when there is, over a period of time, no significant difference between conditions in the patch and in adjacent sediments of similar habitat

Recovery traits Traits that control recovery rates

Remineralised/remineralisation Breakdown of organic material into inorganic components

Resilience The capacity of ecosystems to maintain their functional state in the face of stressors

Response traits Traits that make a species sensitive to a stressor or a disturbance

Scale The window within which we make observations. It encompasses space, time and, for biology, change in the environment and the levels of biological organisation. For example, we can make observations at a site sized 100 m², over a tidal cycle and a mud–sand gradient. Many processes change in importance dependent on the scale at which they are observed.

Scale has different aspects: extent—the total size of the window; lag—the interval or space between sampling points; and grain—the resolution/size of the sample

Seafloor mosaics The spatial arrangement of patchiness (patches within patches) of resources or organisms in or on the seafloor

Seasonality Within-year changes in species abundances and community structure; also the importance of ecological processes often associated with temperature, light or oceanographic variation

Skimming flow Flowing water skims over the surfaces of structures which are not far enough apart for turbulent flow to reach the seafloor

Species occurrence distributions (SOD) Species are sorted into the number of locations that a species is found at

Species sorting Community assemblage is regulated by the response of individual species to the surrounding environment

Species turnover Species changes in occurrence over gradients in space or in time

Structural heterogeneity The number and spatial arrangement of 3-dimensional structures seen across the sediment surface

Succession (in the context of disturbance) The progressive change in community structure and function over time. Often categorised into stages, starting with pioneering or opportunistic species and ending with climax or late successional stage community

Surrogate measures A surrogate is a proxy measure for an attribute of true interest that is too difficult or costly to measure directly

Synergistic The effect of two or more stressors is more than the effect of each singly

Tipping points Tipping points, thresholds, state changes or regime shifts are all non-linear changes over time, driven by complex dynamics

Upwelling Wind-driven and/or topographic-induced motion of dense, cooler and usually nutrient-rich water towards the ocean surface

Weak interactions A community can be structured by strong interactions with key species or by a larger number of weakly interacting species

Index

Tables, figures, and boxes are indicated by an italic *t*, *f*, and *b* following the page number.